walkermaths 3.10

STATISTICAL INFERENCE

NCEA Level 3 Internal

Charlotte Walker and Victoria Walker

Walker Maths 3.10 Statistical Inference
1st Edition
Charlotte Walker
Victoria Walker

Designer: Cheryl Smith, Macarn Design
Production controller: Siew Han Ong

Any URLs contained in this publication were checked for currency during the production process. Note, however, that the publisher cannot vouch for the ongoing currency of URLs.

Acknowledgements
Cover photo courtesy of Shutterstock.

We wish to thank the Boards of Trustees of Darfield and Riccarton High Schools for allowing us to use materials and ideas developed while teaching. Our thanks also go to all past and present colleagues, especially Kath Wilson, who have generously shared their experience and ideas.

For product information and technology assistance,
in Australia call **1300 790 853**;
in New Zealand call **0800 449 725**

For permission to use material from this text or product, please email
aust.permissions@cengage.com

National Library of New Zealand Cataloguing-in-Publication Data
A catalogue record for this book is available from the National Library of New Zealand.

978 0 17 042571 1

Cengage Learning Australia
Level 7, 80 Dorcas Street
South Melbourne, Victoria Australia 3205

Cengage Learning New Zealand
Unit 4B Rosedale Office Park
331 Rosedale Road, Albany, North Shore 0632, NZ

For learning solutions, visit **cengage.co.nz**

Printed in China by 1010 Printing International Ltd
9 25

CONTENTS

Glossary

Make your own glossary of key terms:

Term	Definition	Picture/Example
Discrete variable		
Continuous variable		
Descriptive variable		
Parameters		
Statistics		
Sample		
Population		
Inference		
Sampling variability		

ISBN: 9780170425711

Variability of estimates		
Resampling		
Confidence interval		
Bootstrapped interval		

Bootstrapping

Bootstrapping is the process of:

- obtaining a sample
- resampling from that sample many times
- making an inference about the population.

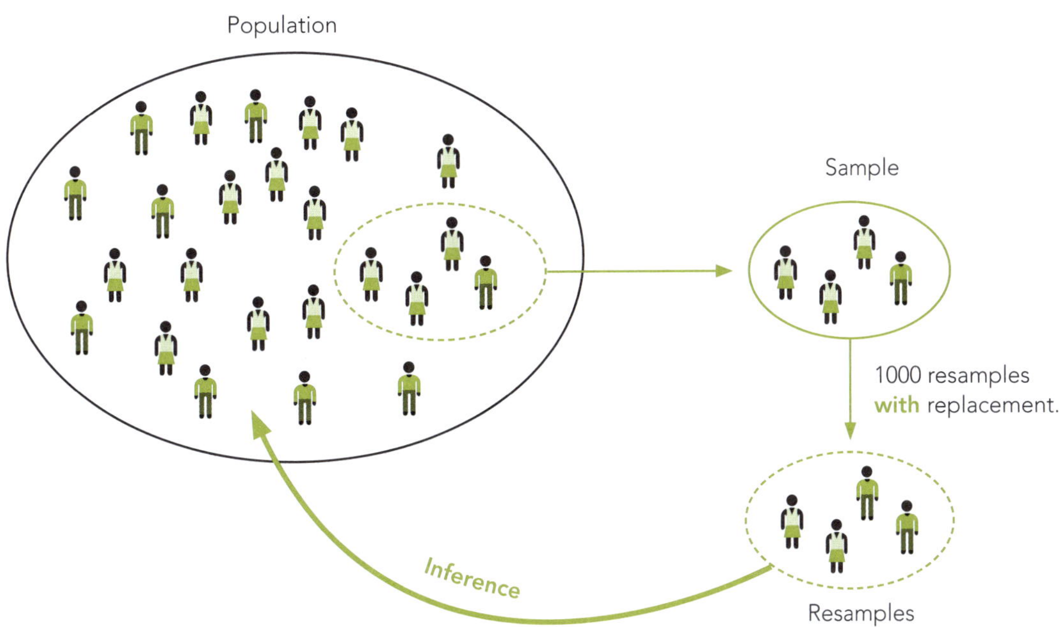

Resampling is **with** replacement, so some individuals could appear several times in a resample, and some may not be in it at all.

ISBN: 9780170425711

Introduction

This standard will require you to investigate data using the statistical enquiry cycle. More specifically you will be required to:

- pose an appropriate question from a multivariate data set for a population
- select and use appropriate displays and measures for the sample
- discuss and compare sample distributions
- make an appropriate inference
- discuss sampling variability, including the variability of estimates
- communicate findings in a conclusion.

Statistical enquiry cycle

Your report should connect back to all aspects of the statistical enquiry cycle:

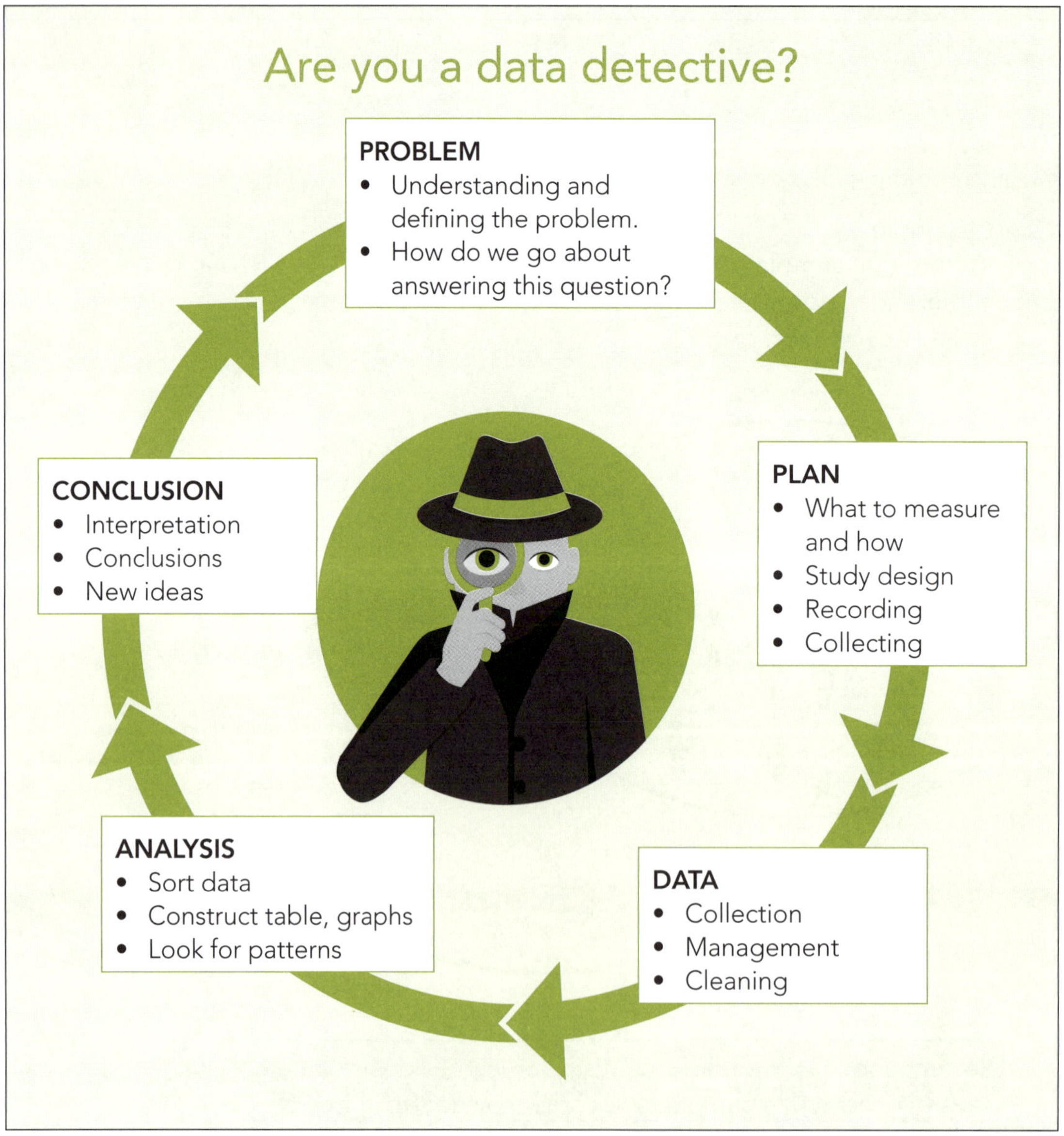

Are you a data detective?

PROBLEM
- Understanding and defining the problem.
- How do we go about answering this question?

PLAN
- What to measure and how
- Study design
- Recording
- Collecting

DATA
- Collection
- Management
- Cleaning

ANALYSIS
- Sort data
- Construct table, graphs
- Look for patterns

CONCLUSION
- Interpretation
- Conclusions
- New ideas

 ISBN: 9780170425711

Revision

Types of variables

There are three types of variables:

Continuous variables These are measured values, which can be **fractions** or **decimals**. Examples: height, weight, distance, time.

Discrete variables These are **counted** values, which are usually **whole** numbers. Examples: number of pets, shoe size (although this is unusual because these can be in halves).

Descriptive variables These are **categorical** variables, which are usually **words**. Example: eye colour, type of pet.

In theory we usually analyse **continuous** data. However, in practice, provided there is a large enough range, we could analyse **discrete** data.

Complete the table by stating what type of variable each is, and whether or not (✓ or ✗) each would be a suitable variable to use as part of an inference study.

Variable	Type of variable	Suitable for inference?
Favourite ice-cream flavour		
Height		
Number of apps on your phone		
Neck circumference		
Shoe size		
Number of siblings		
Size of your house (m^2)		
Number of countries you have visited		
Hair colour		
Brand of shoe		
Distance to the nearest supermarket		
Length of longest finger		
Weight of school bag		
Amount of water you drink today		
Age		

In order to make an inference about the population, we usually compare two sets of **continuous** data; for example, the heights of males and the heights of females.

ISBN: 9780170425711

Box plots

1 Complete this diagram by using the words in the box.

Median Maximum Upper quartile Minimum Lower quartile

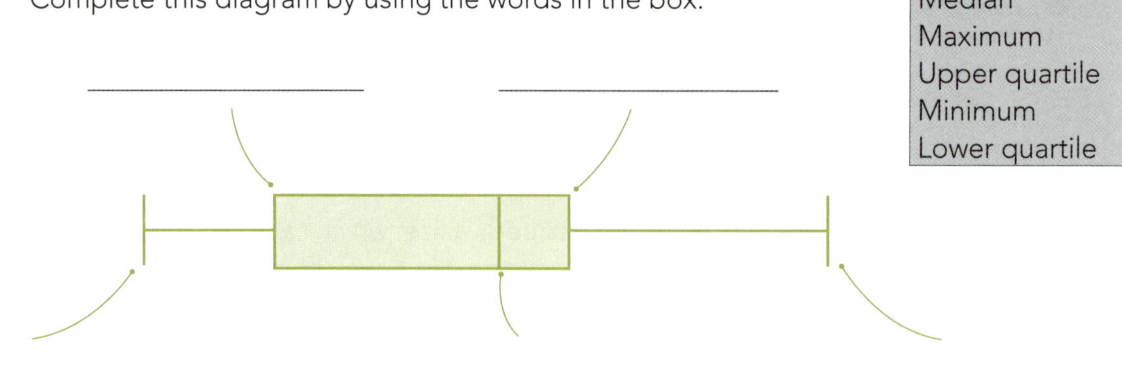

2 Find the appropriate statistics and draw the box plots for these data sets.

Number of text messages sent between 3pm and 4pm from their phones.

Boys: 6, 21, 19, 14, 13, 19, 16, 5, 9, 12, 15, 10, 13, 13, 17, 14, 6, 10, 20

Girls: 10, 7, 13, 3, 18, 23, 6, 10, 17, 19, 8, 7, 9, 16, 21, 22, 20, 17

Boys:
Minimum ____ Lower quartile ____ Median ____ Upper quartile ____ Maximum ____

Girls:
Minimum ____ Lower quartile ____ Median ____ Upper quartile ____ Maximum ____

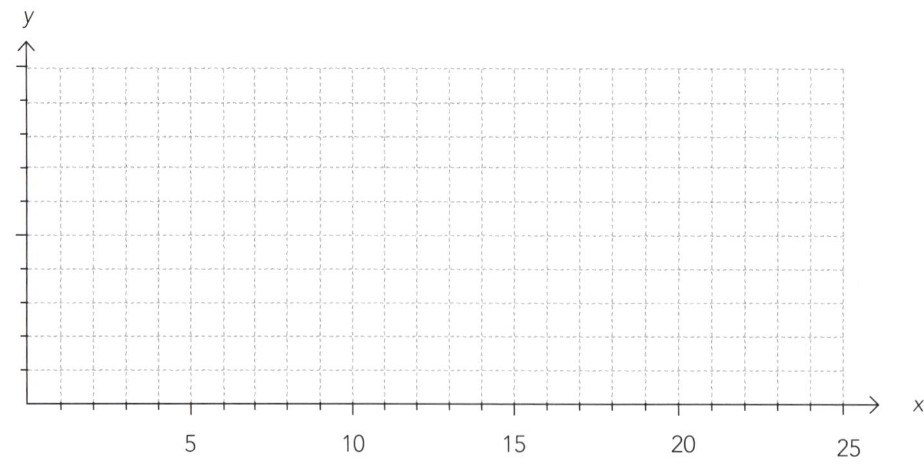

ISBN: 9780170425711

3 Write in the number of pieces of data that is represented by each interval.

Sample size = 24

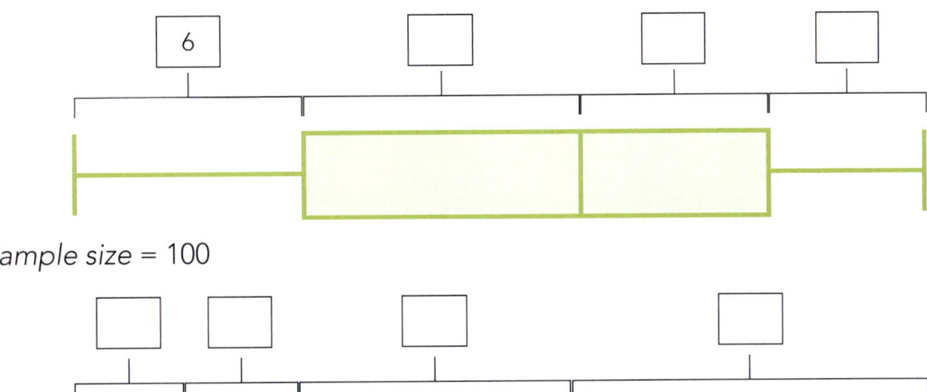

Sample size = 100

4 Fill in the missing percentages.

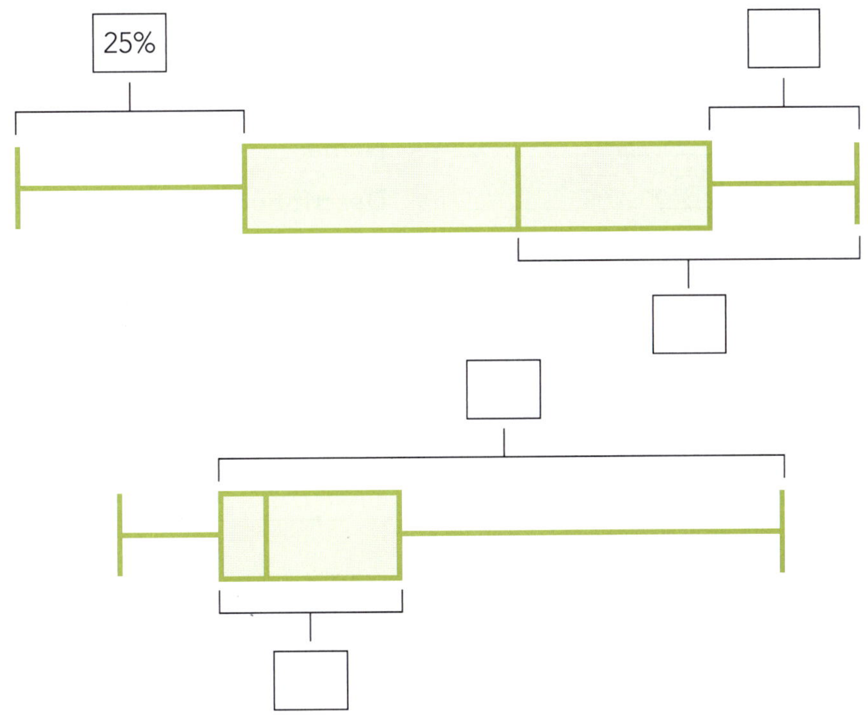

5 Write in the boxes the number of pieces of data that are represented by each interval.

Sample size = 216

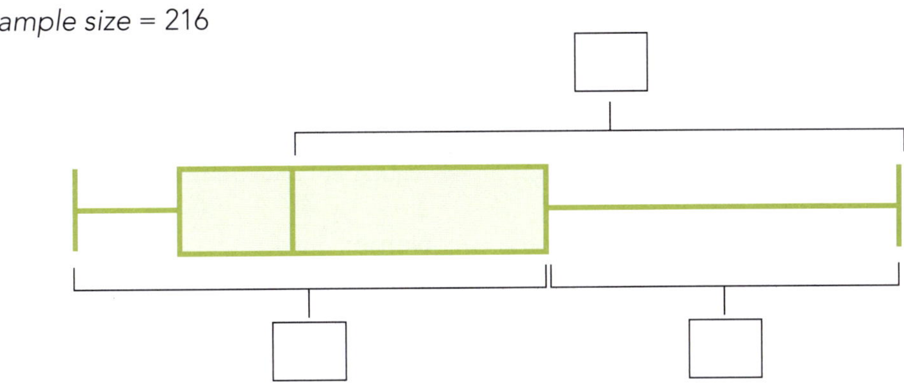

Shapes of distributions

Match the pictures with the names and descriptions of the distributions.

Names:

Triangular | Right skew | Uniform
Bimodal | Irregular | Left skew

Descriptions:

- Symmetrical and bell-shaped curve.
- The tail is on the right-hand side.
- Symmetrical and trianglar shape.
- No real pattern.
- Looks like a box.
- There are two peaks.

1

Name: Bell shaped distribution

Description: ______

2

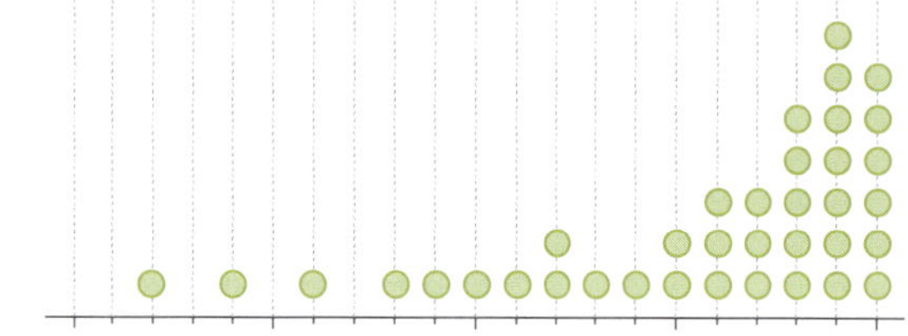

Name: ______

Description: The tail is on the left-hand side.

3

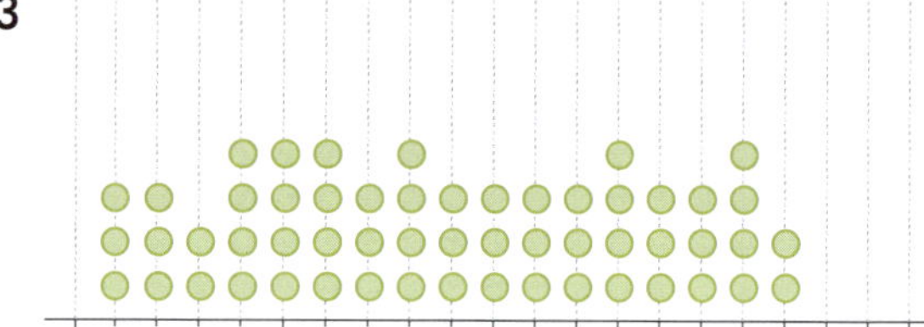

Name: ______

Description: ______

4

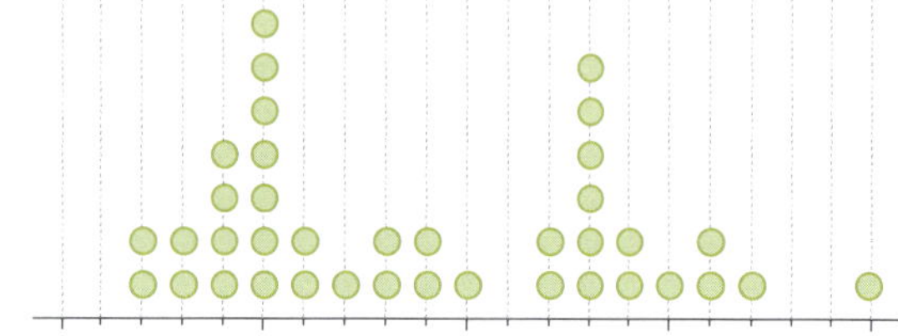

Name: ______

Description: ______

5

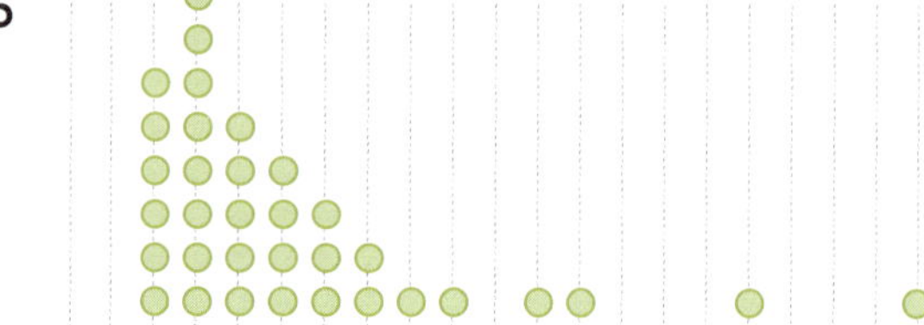

Name: ______

Description: ______

6

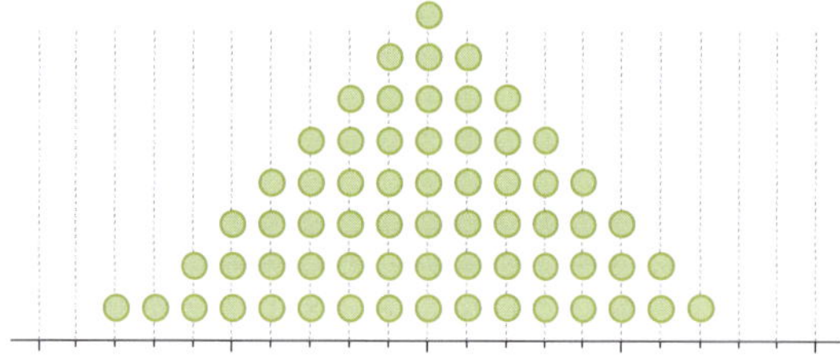

Name: ______

Description: ______

7

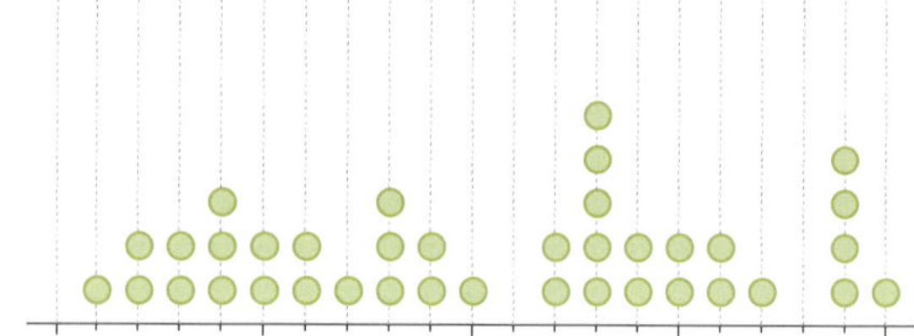

Name: ______

Description: ______

 ISBN: 9780170425711

Use of medians and means

- You will be required to use either the **median** or the **mean** as your measure of central tendency.
- Justification for the use of either can add depth to your report.

Median = the middle number (or the mean of the middle two numbers). Remember that to find this, the numbers need to be put in order.

Mean = the sum of all the data values divided by the number of pieces of data.

$$\text{Mean} = \frac{\text{sum of all values}}{\text{number of pieces of data}}$$

- For **symmetrical** distributions, both **mean** and **median** are appropriate.
- For distributions that are **not symmetrical**, the **median** should be used because it is not influenced by extreme values.

Example: A restaurant owner asked nine customers to fill in a customer satisfaction survey on a scale of 1 to 10. These are the results:

1, 2, 3, 3, 3, 5, 9, 10, 10

Median = 3

$$\text{Mean} = \frac{1 + 2 + 3 + 3 + 3 + 5 + 9 + 10 + 10}{9}$$

$$= 5.11$$

If the owner used the **median**, she might conclude that customers are not very satisfied. If the owner used the **mean**, she might conclude that the customers are reasonably satisfied.
In this situation, because the majority of customers scored the restaurant under 5, it would be more appropriate to use the median as the average score for the satisfaction survey.

The mean may not always be affected by unusual points. The sample statistics should be investigated before making this call.

For these data sets, calculate the median and the mean and identify which would be more appropriate to use and why.

1 Calculate the median and the mean for the following data sets.

a

3, 5, 7, 7, 8, 10, 12, 15, 29

Median = ____________ Mean = ____________

It would be appropriate to use **the median/the mean/either the median or mean**

because ____

____

ISBN: 9780170425711

b 12, 14, 15, 17, 18, 20 20, 26, 27, 31

Median = ______________ Mean = ______________

It would be appropriate to use **the median/the mean/either the median or mean** because ____

____

c 5, 3, 12, 6, 4, 9, 9, 10, 40, 12, 7, 16, 9

Median = ______________ Mean = ______________

It would be appropriate to use **the median/the mean/either the median or mean** because ____

____

2 These are the number of Level 3 NCEA credits that a sample of students earned by the end of the first term.

Boys: 0, 0, 3, 4, 4, 5, 6, 8, 9, 9, 9, 12, 12, 15, 16, 25

Median = ______________ Mean = ______________

Girls: 0, 0, 0, 0, 2, 4, 4, 4, 8, 9, 9, 12, 12, 12, 14

Median = ______________ Mean = ______________

Create two dot plots of this data.

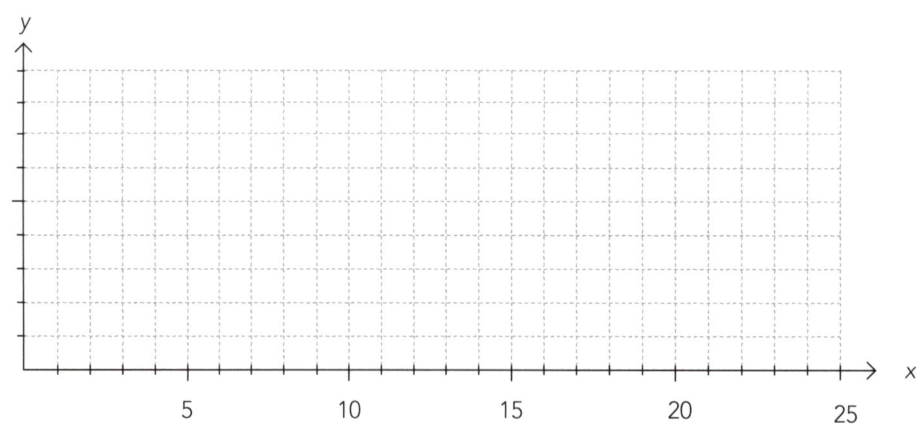

For the boys' data, the **median/mean** is appropriate to use because:

____

____

For the girls' data, the **median/mean** is appropriate to use because:

____

____

 ISBN: 9780170425711

3 Identify whether the median or the mean or either would be appropriate. Justify your answer.

a

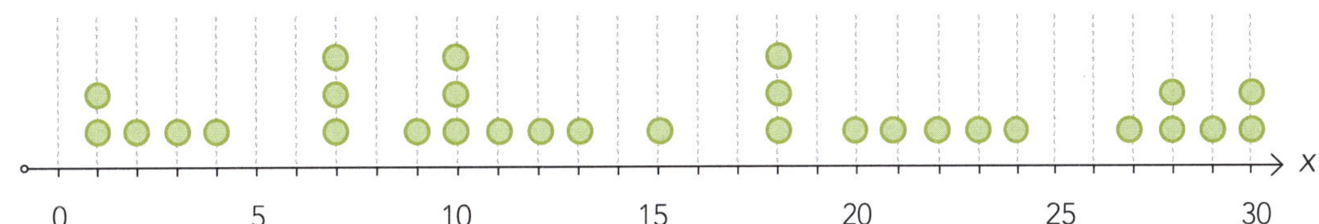

The **median/mean/median or mean** would be suitable because the dot plot is **symmetrical/asymmetrical**.

b

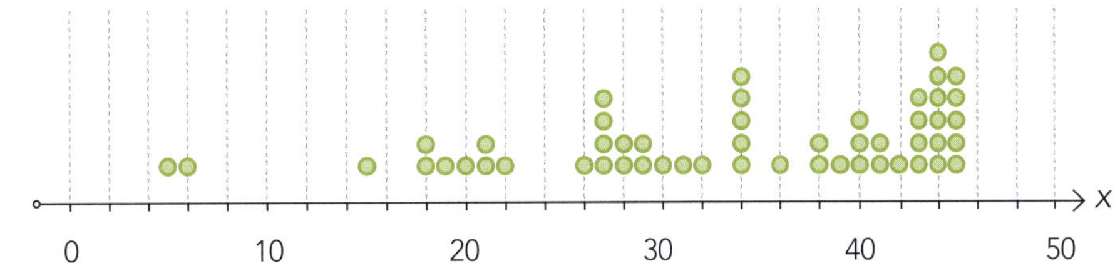

The **median/mean/median or mean** would be suitable because the dot plot is ______________________________.

c

The **median/mean/median or mean** would be suitable because the box plot is ______________.

d

The **median/mean/median or mean** would be suitable because the box plot is ______________.

e

The **median/mean/median or mean** would be suitable because the box plot is ______________.

f Appropriate measure(s) of centre for a symmetrical distribution:

____

g Appropriate measure(s) of centre for an asymmetrical distribution:

____

ISBN: 9780170425711

Finding random numbers on a scientific calculator

Step 1: Find the RAN or RAN# button.

Try pressing the = button. You will get a four-digit number between 0.001 and 0.999. You can use the digits after the decimal points as your random numbers, but it is easier if you follow **steps 2** and **3** below.

Note: If you use the four-digit numbers, beware numbers with fewer digits — your calculator will have left off the final zeros.

Step 2: If you need 8 random numbers ranging from 1 to 6, enter:

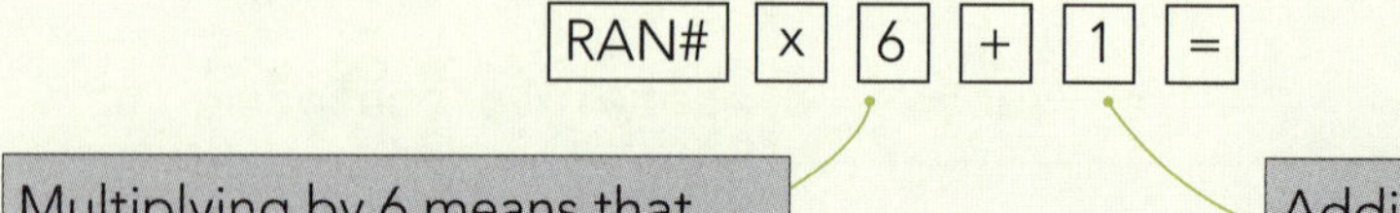

Multiplying by 6 means that the random numbers all lie between 0.006 and 5.994.

Adding 1 means that the random numbers all lie between 1.006 and 6.994.

Step 3: Press = 8 times and record each number.

Example of results:

Use the digit before the decimal point as your random number (do not round).

On calculator	5.464	6.034	1.864	1.64	2.626	3.364	2.278	6.832
Random number	5	6	1	1	2	3	2	6

Another example:

Step 2: If you need 7 random numbers ranging from 1 to 58, enter:

Multiplying by 58 means that the random numbers all lie between 0.058 and 57.942.

Adding 1 means that the random numbers all lie between 1.058 and 58.942.

Step 3: Press = 7 times and record each number.

Example of results:

On calculator	36.496	21.996	12.832	11.15	36.438	54.824	7.728
Random numbers	36	21	12	11	36	54	7

ISBN: 9780170425711

Finding random numbers on a graphics calculator

Step 1: ⇒ RUN
⇒ EXE
⇒ OPTN
⇒ F6 ▷
⇒ F3 PROB
⇒ F4 RAND
⇒ F2 Int

Step 2: If you need 4 random numbers ranging from 1 to 150, enter:

If you have an older graphics calculator, follow the first five instructions from Step 1 and then enter:

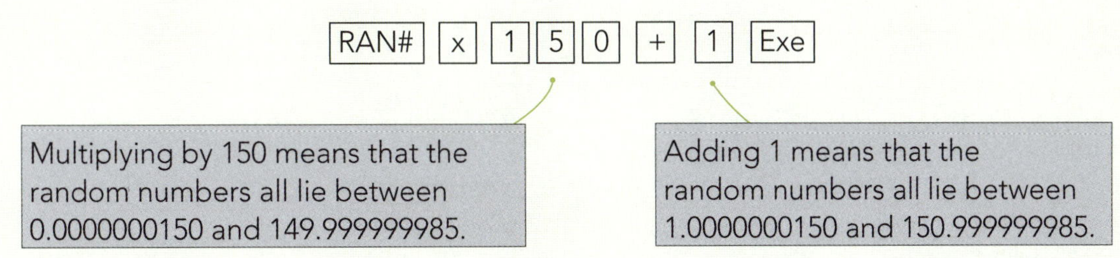

Step 3: Press Exe 4 times and record each number.

Example of results:

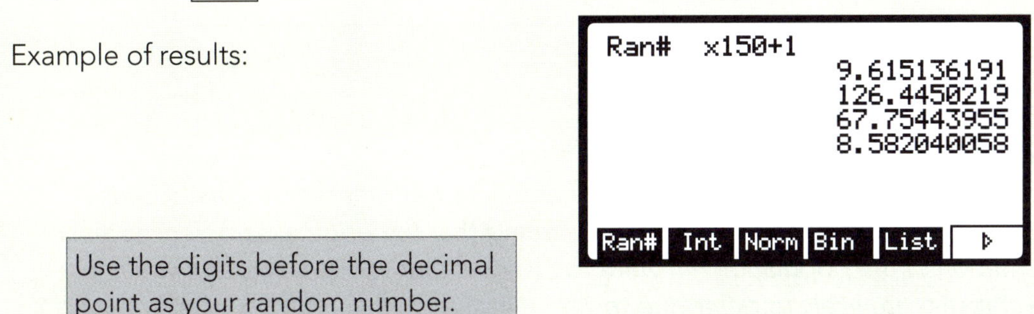

On calculator	9.615136191	126.4450219	67.75443955	8.582040058
Random numbers	9	126	67	8

Note: In practice you will be finding many more random numbers than in these examples.

ISBN: 9780170425711

Sampling variability

- Different samples from the same set of data will almost certainly produce **different** results.
- This will be the case whether sampling with or without replacement.
- This is known as **sampling variability**.

These are samples from the same population:

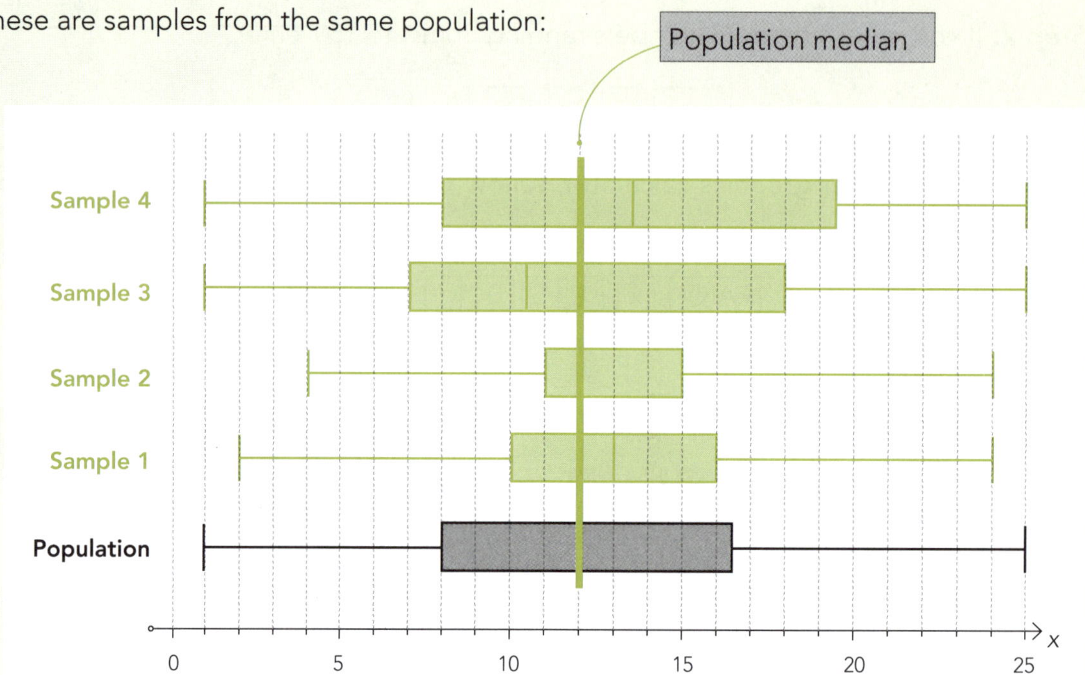

Example of how this could be discussed:
If we took another sample we would, almost certainly, get different results because the sample data would be different. I am assuming that the original sample was a fair representation of the population. If I were to take another sample, the displays and sample statistics are likely to differ due to sampling variability.

Remember: It is important that every piece of data has the same chance of being selected for the sample. If it does not, then the sample will be **biased**.

Example of bias
If researchers measure only the kea that come close enough to be hand fed, this would result in a biased sample. Kea that live in more populated areas are more likely to be used to being fed, which means they are more likely to be selected for measuring. It wouldn't be surprising if these kea were heavier!

ISBN: 9780170425711

1 Here are the weights of the kea that live on an isolated island (i.e. a population). Randomly select a sample of nine kea and then calculate the median. Do this **without** replacement (i.e. you should not have the same kea twice in a sample). Note: For this activity, you could use the data cards in the middle of this book or random numbers.

Random number	Kea	Mass (g)
17	Taan	1000
14	Moritz	900
27	Aroha	700
12	Anthea	750
9	Hoheria	
5		
15		
7		
18		

Median = ________

Finish this table and calculate the median.

Random number	Kea	Mass (g)
9		

Median = ________

Random number	Kea	Mass (g)

Median = ________

	Kea name	Mass (g)
1	Templeton	810
2	Sneaky	840
3	Ejay	750
4	Mungbean	700
5	Lorraine	850
6	Christmas	850
7	Sage	950
8	Kellie	760
9	Hoheria	780
10	Pong	580
11	Camel	580
12	Anthea	750
13	Monty	810
14	Moritz	900
15	Marie	770
16	Ping	740
17	Taan	1000
18	Alwyn	800
19	Yin	680
20	Arya	925
21	Zeal	650
22	Swazi	675
23	Gizo	860
24	Roxanne	690
25	Anna	730
26	Juno	900
27	Aroha	700
28	Bud	950
29	Tool	770
30	Handlebars	915

What do you notice about the medians?

____

____

____

____

ISBN: 9780170425711

2 These are the total number of Level 3 credits for Year 13 students at the end of term 1 from Paradise High School (i.e. a population).

Using random numbers to select three samples of 10 students, display the data in a dot plot, then calculate the median. Do this **without** replacement (i.e. you should not have the same student twice in a sample).

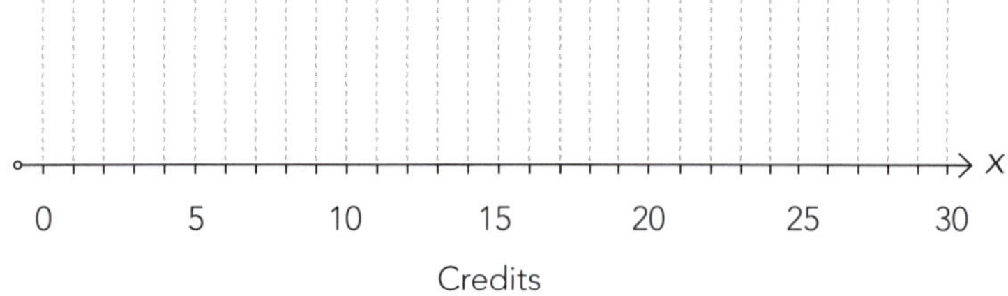

Median = ______________

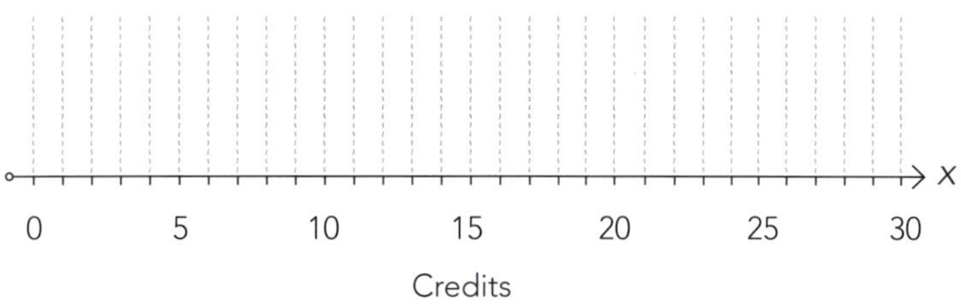

Median = ______________

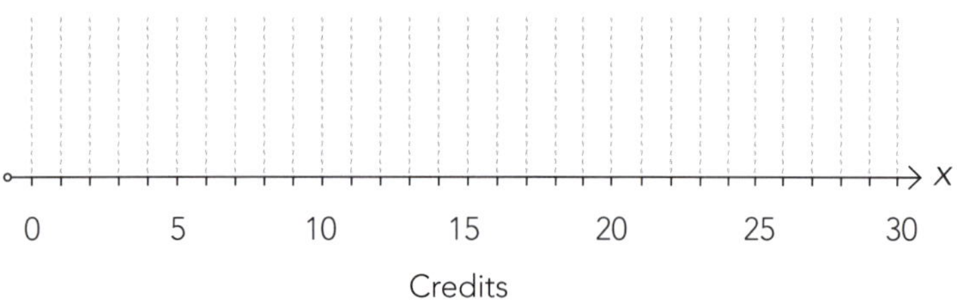

Median = ______________

What do you notice about the medians?

____

____

____

____

____

Number	Name	Number of credits
1	Caleb	4
2	Emma	7
3	Ethan	6
4	Amber	0
5	Jonathan	6
6	Jacob	4
7	Hunter	0
8	Joshua	3
9	Shae	2
10	Luke	0
11	William	8
12	Ryan	7
13	Zara	4
14	Harry	4
15	Amelia	7
16	Katherine	0
17	Olivia	4
18	Georgia	4
19	Joseph	4
20	Toni	0
21	George	2
22	James	6
23	Courtney	0
24	Adrian	4
25	Campbell	4
26	Lucy	4
27	Julia	11
28	MacKenzie	4
29	Eron	3
30	Imogen	30
31	Teri	7
32	Taylor	0

ISBN: 9780170425711

Resampling

- Sometimes, due to expense, availability, etc., it is only possible to obtain a **small sample** from a population.
- It is unlikely that the sample median is the same as the population median, particularly if the sample is small.
- In order to make an inference about the interval within which the population median is likely to lie, **many resamples (with replacement)** are taken from the **sample**.
- The medians (or means) of these form the basis for our **bootstrapped inference**.
- The bootstrapped inference will be reliable only if the original sample is unbiased and representative of the population.

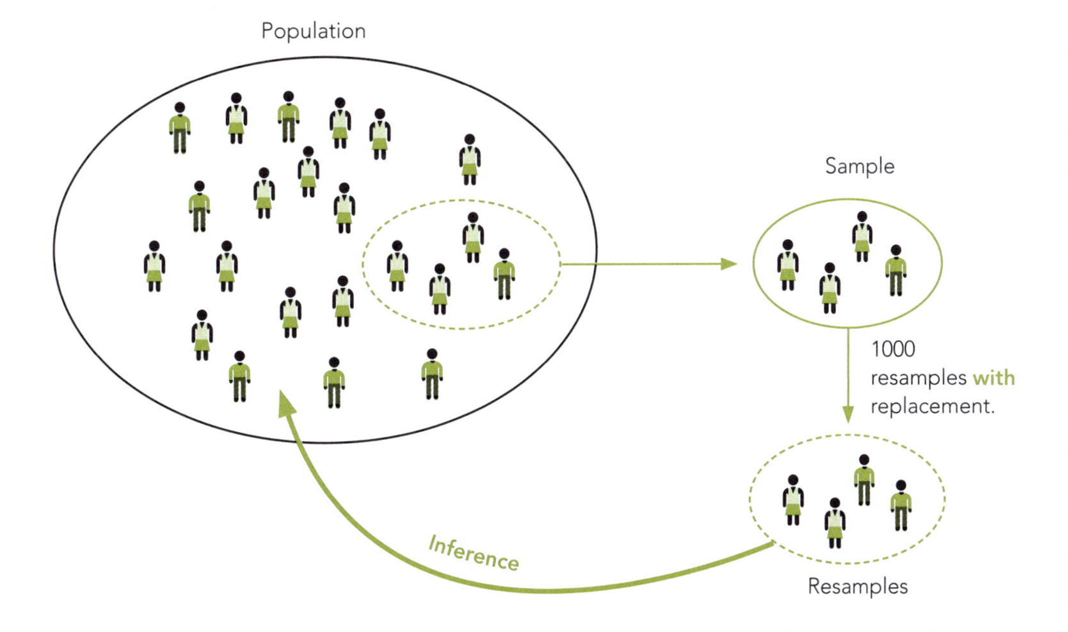

1 Using the sample of six adult kea, complete the resampling and calculations below.

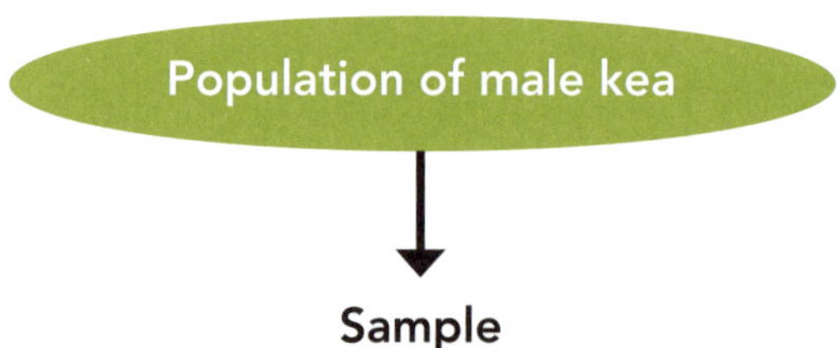

Sample

Number	Kea ID	Mass (g)
1	Daffy	860
2	Cheezel	930
3	Herbert	840
4	Baggins	820
5	Bandit	880
6	Allsorts	1010

Median = 870 g

I rolled a die and got 3, 6, 3, 5, 4, 2.

Resample 1	
Number	**Mass (g)**
3	840
6	1010
3	840
5	880
4	820
2	930

Median = 860 g

Resample 2	
Number	**Mass (g)**
5	
6	
6	
1	
4	
6	

Median = ____________

Resample 3	
Number	**Mass (g)**

Median = ____________

Notice that some kea were selected more than once and some not at all.

Complete these two resamples and calculate the medians.

Note that:
- We are likely to get **different resamples** each time, and **different medians**.
- Resamples contain the same number of subjects as in the original sample. For example we had six kea so we resampled six.

ISBN: 9780170425711

2 Below are the arm spans of students from Paradise High School. Resample this set of data.

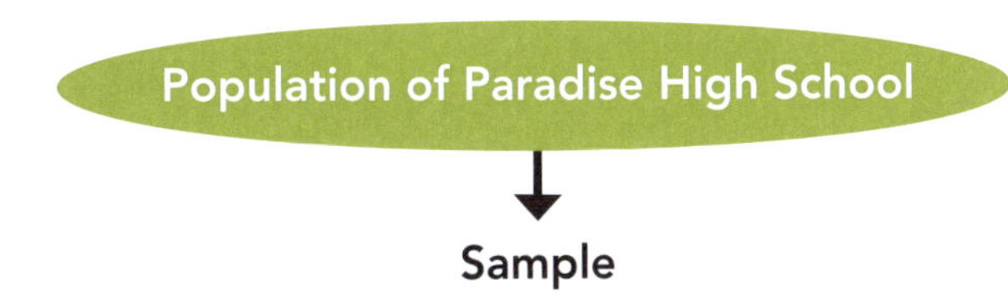

Sample

Number	Student	Arm span (cm)
1	Bruce	180
2	James	175
3	Nick	170
4	Hemi	187
5	Trevor	162
6	Jai	174
7	Luther	173
8	Ravji	185
9	Spencer	153
10	Ollie	181

You will need to find a way to randomly sample these 10 students, e.g. cards, random number generator, etc.

Median = ____________

Resample 1

Number	Arm span (cm)

Median = ____________

Resample 2

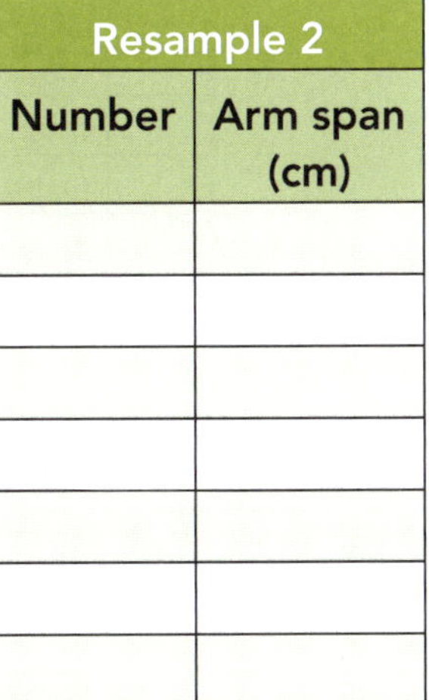

Number	Arm span (cm)

Median = ____________

Resample 3

Number	Arm span (cm)

Median = ____________

Resample 4

Number	Arm span (cm)

Median = ____________

Note that:

- In order to use this method in practice, a **very large number of resamples** need to be taken.
- This would be very time-consuming. Therefore we have computer programs which can do this very rapidly.
- Usually 1000 resamples are taken.

ISBN: 9780170425711

The difference between two medians (or means)

- We often need to be able to compare medians (or means) for two groups within the population.
- The computer program will take **1000 pairs of resamples**.
- It will create a **dot plot** showing the difference between medians (or means) for each pair.
- It will calculate the interval within which the central **950** differences lie.
- This interval is called the **95% confidence interval** for the difference between two medians (or means).

We can be 95% confident that this interval will contain the difference between the medians (or means) for the population.

Note: For the resamples, we always subtract the median of the group with the smaller sample median from the group with the larger sample median.

Therefore it is possible to get **negative** values when calculating the difference between resampled medians or means.

1 Below is a sample of arm spans of Year 13 students from Paradise High School.

Samples

Males

Number	Student	Arm span (cm)
1	Bruce	180
2	James	175
3	Spencer	153
4	Hemi	187
5	Trevor	162

Median = ___________

Females

Number	Student	Arm span (cm)
1	Maya	165
2	Charlie	154
3	Elaine	162
4	Jane	171
5	Eleanor	167

Median = ___________

Which group has the larger median? ___________

What does this mean when calculating the difference in the resamples? ___________

____

____

Resample 1

Complete the resampling that has been started for you

Males

Number	Student	Arm span (cm)
1		
1		
4		
2		
1		

Median = ___________

Females

Number	Student	Arm span (cm)
3		
5		
2		
1		
3		

Median = ___________

 ISBN: 9780170425711

Calculate the difference in the medians. __________ – __________ = __________

What does this mean for this resample?
In this resample the difference between the medians was **positive/negative**. This means that in this resample, the median arm span of the **males/females** was longer than that of the **males/females**.

Resample 2

Males

Number	Student	Arm span (cm)
5		
3		
1		
3		
4		

Median = __________

Females

Number	Student	Arm span (cm)
1		
2		
1		
5		
4		

Median = __________

Calculate the difference in the medians. __________ – __________ = __________

What does this mean for this resample?
In this resample the difference between the medians was **positive/negative**. This means that in this resample, the median arm span of the **males/females** was longer than that of the **males/females**.

Resample 3

Males

Number	Student	Arm span (cm)
5		
4		
2		
5		
3		

Median = __________

Females

Number	Student	Arm span (cm)
2		
5		
3		
4		
2		

Median = __________

Calculate the difference in the medians. __________ – __________ = __________

What does this mean for this resample?

In this resample the difference between the medians was ______________. This means that in this resample, the median arm span ____

____.

ISBN: 9780170425711

2 Below are samples of kea. Resample this set of data.
Note: For this activity, you could use the data cards in the middle of this book.

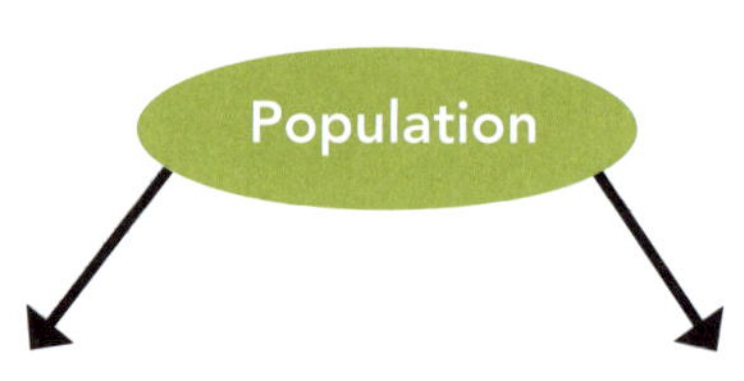

Sample of male kea

Number	Kea ID	Mass (g)
1	Daffy	860
2	Cheezel	930
3	Herbert	840
4	Baggins	820
5	Bandit	880
6	Allsorts	1010

Median = ____________

Sample of female kea

Number	Kea ID	Mass (g)
1	Beryl	840
2	Xena	830
3	Cassidy	790
4	Creeky	868
5	Trixy	900
6	Sage	950

Median = ____________

In the resamples you will need to subtract the **male/female** medians from the **male/female** medians.

Resample 1

Males

Random number	Kea ID	Mass (g)
4	Baggins	820
5	Bandit	880
4	Baggins	820
1	Daffy	860
1	Daffy	860
5	Bandit	880

Median = 860 g

Complete this resample and then complete another three resamples for each sex.

Females

Random number	Kea ID	Mass (g)
3	Cassidy	790
4		
1		
6		
1		
6		

Median = ____________

Difference between the medians = ______________

Resample 2

Males

Random number	Kea ID	Mass (g)

Median = ____________

Females

Random number	Kea ID	Mass (g)

Median = ____________

Difference between the medians = ______________

ISBN: 9780170425711

Resample 3

Males

Random number	Kea ID	Mass (g)

Median = ___________

Females

Random number	Kea ID	Mass (g)

Median = ___________

Difference between the medians = _______________

Resample 4

Males

Random number	Kea ID	Mass (g)

Median = ___________

Females

Random number	Kea ID	Mass (g)

Median = ___________

Difference between the medians = _______________

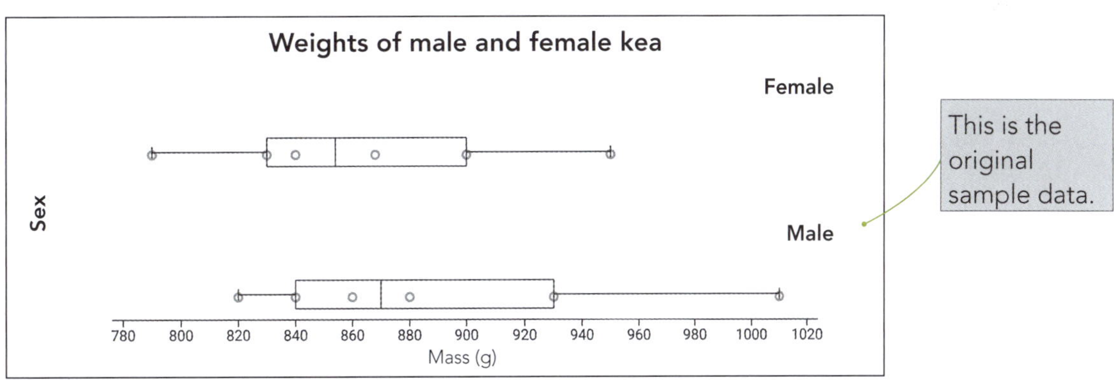

Plot the difference between the medians on a dot plot (the first one has been done for you).

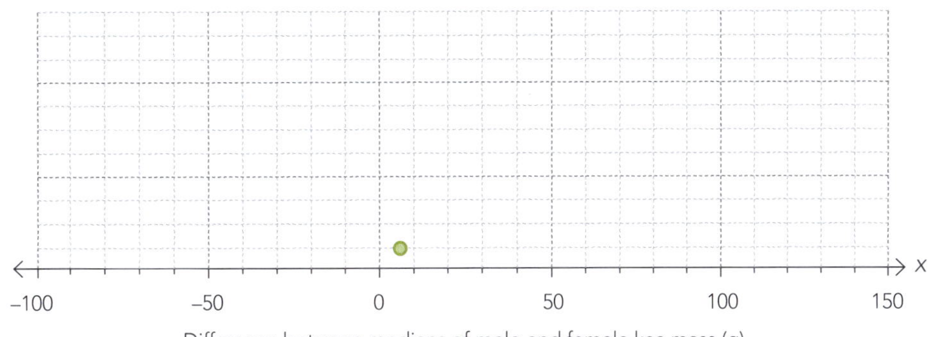

Your computer program will do this 1000 times for you.

ISBN: 9780170425711

Computer program output

- Your graphing program will produce:
  1 dot and box plots for the **sample** data for each of the two groups
  2 a dot plot showing the differences between the medians (or means) of each **resample**, along with:
    a the difference between the medians of the two samples
    b the boundary values for the central 95% of the differences between the medians of the resamples.
- You may have to add or modify:
  1 the title
  2 the axis labels, including units.

Examples:
This is the output you would see if you were looking at the difference between the **medians**:

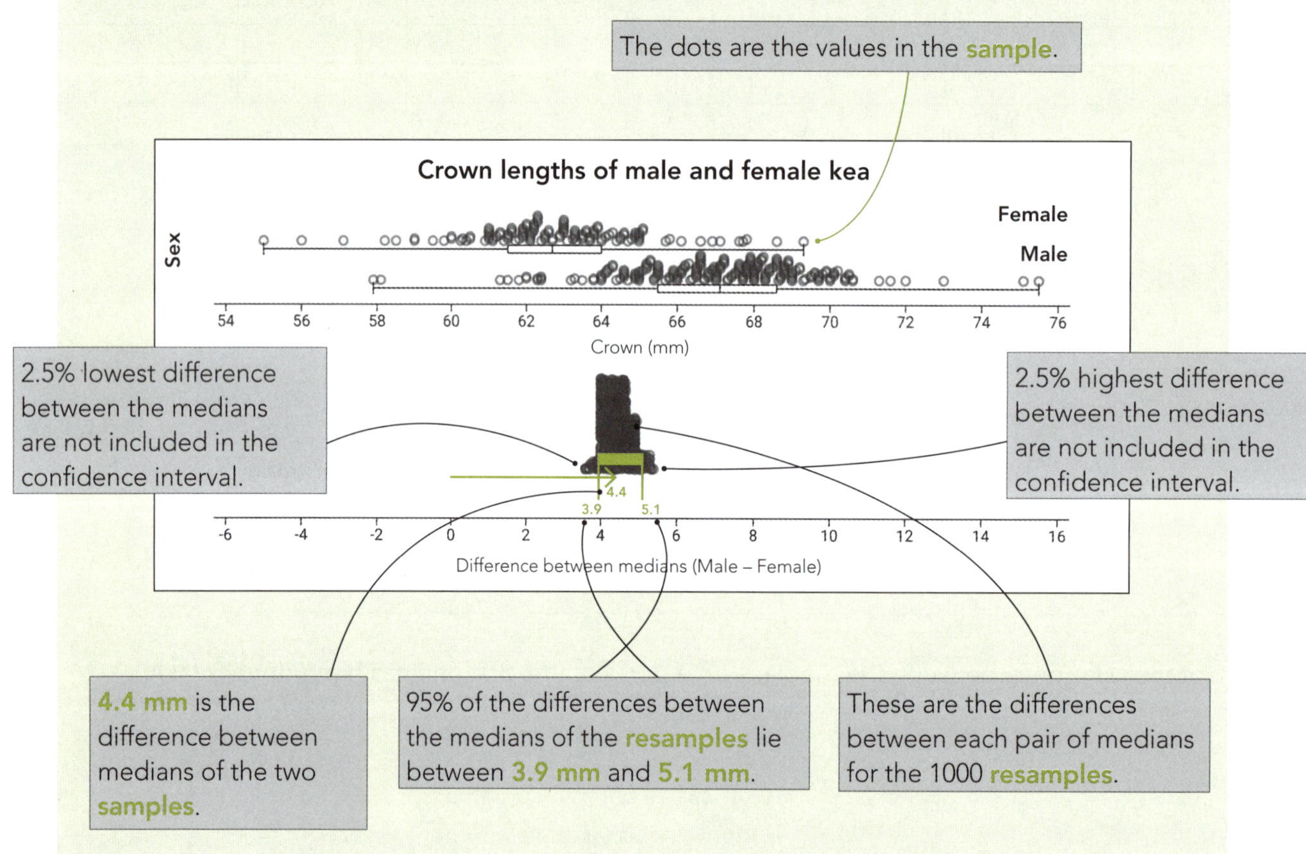

  ISBN: 9780170425711

This is the output you would see if you were looking at the difference between the **means**:

4.24 mm is the mean difference of your two **samples**.

The hollow dots are the values in the sample. The **solid green dots** are the **sample** means.

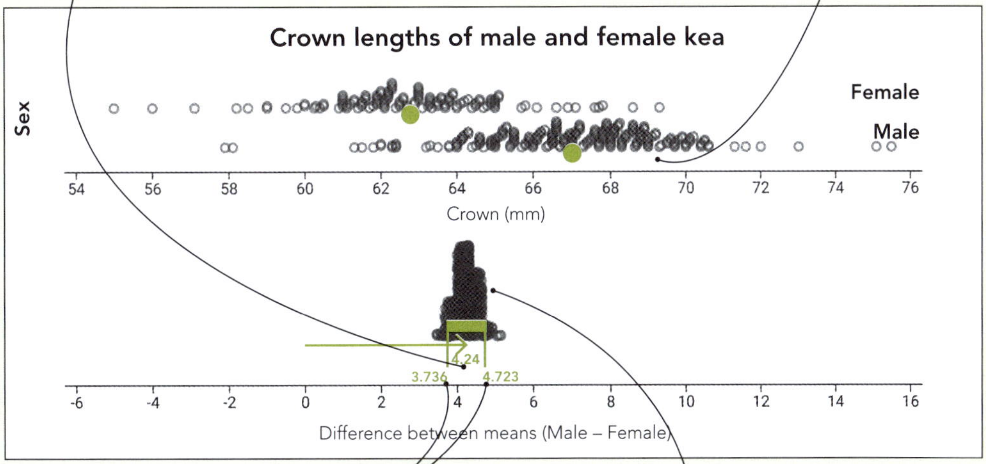

95% of the differences between the means of the **resamples** lie between **3.736 mm** and **4.723 mm**.

These are the differences between each pair of means for the 1000 **resamples**.

1 Why is there no box plot in this graph?

____

____

____

2 Why is this confidence interval's value different from the one on page 26?

____

____

____

3 Why do we subtract the female kea **median/mean** from the male kea median/mean?

____

____

____

ISBN: 9780170425711

Does sample size make a difference?

Below are bootstrapped intervals calculated from **different sample sizes**.

- Notice that for **larger** samples, the interval gets **smaller**.

Sample of 10 kea

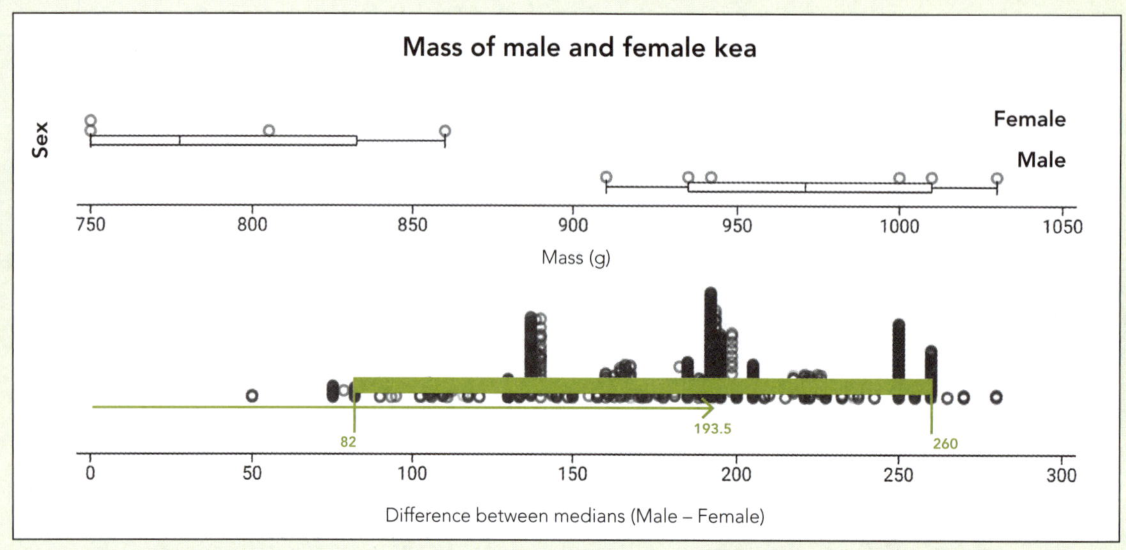

Sample of 40 kea

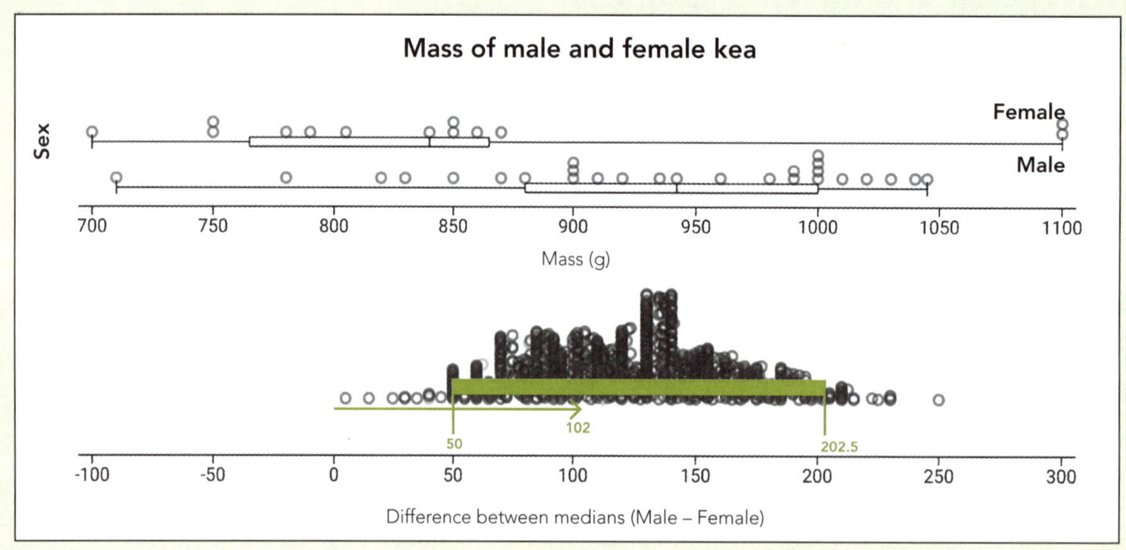

 ISBN: 9780170425711

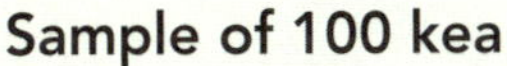

Sample of 100 kea

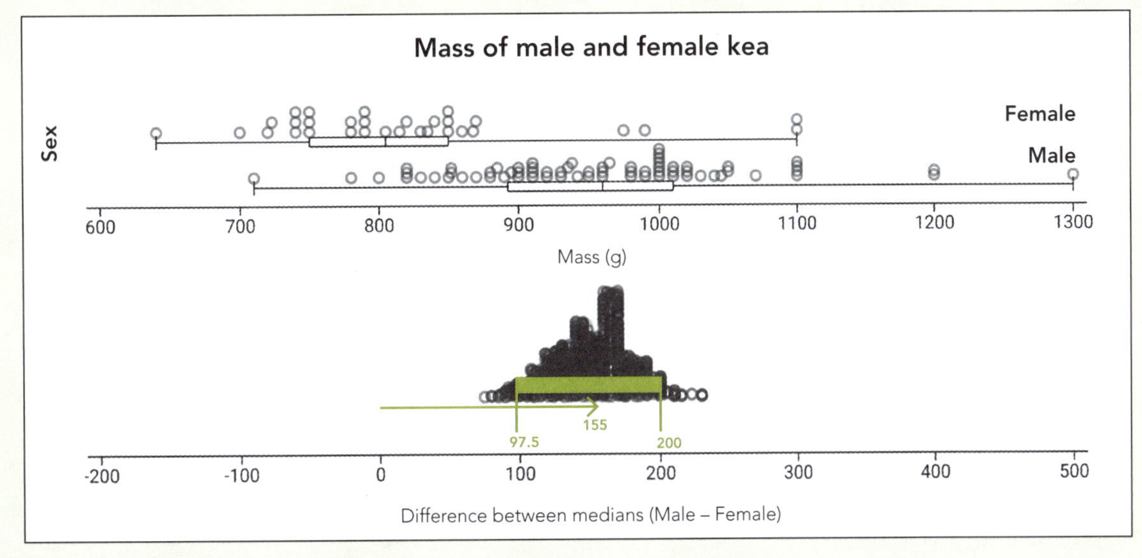

In summary

Complete the table.

Sample size	Upper limit of confidence interval	Lower limit of confidence interval	Difference = width of confidence interval
10	260		178
40		50	
100			

The width of the confidence interval becomes ______________________________ as the sample size ______________________________.

An increased sample size reduces the **range** of the bootstrapping interval. It does NOT change the level of confidence.

ISBN: 9780170425711

Setting up your investigation

You must:
1 Select suitable data.
2 Pose an appropriate question informed by contextual knowledge.
3 Define the variables.
4 State the purpose of the investigation and who would find it useful.

Example: A number of kea have been caught from a variety of locations throughout New Zealand and data has been collected. For each kea, the following are recorded:
- Kea ID
- Weight in grams
- Beak length in mm
- Crown length in mm
- Age: Fledgling = less than 1 year
  Juvenile = 1–2.4 years
  Sub-adult = 2.5–4 years
  Adult = more than 4 years
- Sex.

ISBN: 9780170425711

1 Select suitable data

The data you select needs to be:

- numerical
- two subsets, for example male and female
- interesting to analyse and research.

Kea ID	Mass (g)	Beak length (mm)	Crown (mm)	Age	Sex
Katsu	790	43.8	62.5	Sub-adult	Female
Bluff	738	43.1	61.7	Sub-adult	Female
Norman	980	48.7	65.7	Sub-adult	Male
Juno	800	41.4	62.1	Sub-adult	Female
Frosty	870	42.7	63	Adult	Female
Pavlova	680	39.3	59.3	Juvenile	Female
Daffy	860	46.8	64.15	Adult	Male
Creeky	868	41.5	61.9	Adult	Female
Houdini	800	49.4	64.5	Sub-adult	Male
River	770	43.5	62.3	Sub-adult	Female
Apollo	1000	49	68.8	Juvenile	Male
Cheezel	930	52.2	67.8	Adult	Male
Dante	900	47.4	66	Juvenile	Male
Buddy	860	52	68.5	Juvenile	Male
Manuka	950	48.5	64.6	Adult	Female
Herbert	840	48.6	65.2	Adult	Male
Jolly	1010	48.1	66.9	Juvenile	Male
Sheila	980	42.7	62.5	Adult	Female
Roxette	950	47.1	65.6	Sub-adult	Female
Daisy	580	46.5	61.7	Juvenile	Female
Foxy	688	42.3	62.9	Juvenile	Female
Baggins	820	51.5	68.6	Adult	Male
Hoki	700	39.9	60.8	Sub-adult	Female
Gordon	900	49.2	62.4	Sub-adult	Male
Bruce	820	47.1	65.05	Sub-adult	Male
Olive	800	42	64	Sub-adult	Female

1 Identify some groups and an appropriate numerical value you could investigate from the table above. (The first one has been done for you.)

Group one	Group two	Variable
Male	Female	Mass (g)

2 Could you compare the mass of different age groups of kea? Explain your answer.

____

____

____

ISBN: 9780170425711

2 Pose an appropriate question, using suitable data

Features of an appropriate question:

- It should include details of **units** and **context**.
- It should **not** just have a yes or no answer.
- It should be informed by contextual knowledge (which is likely to require research).

Suggested format:

> **What is the difference between the median/mean (context) of (group one, population) and (group two, population)?**

Examples:

What is the difference between the median of Year 13 boys and Year 13 girls?

Doesn't have the parameter or population in the question.

What is the difference between the median height (cm) of Year 13 boys and Year 13 girls in New Zealand?

Your choice and the quality of your question may affect the potential depth of, and therefore standard of, your final report.

Examples of some good questions:

1 What is the difference between the median **weight (g)** of **male** and **female kea** in **New Zealand**?
2 What is the difference between the median **crown length (mm)** of **juvenile** and **adult kea** in **New Zealand**?

1 State whether the following are good questions or not, and if necessary, rewrite the question.

a What is the difference between the number of sunshine hours in Auckland and Nelson?

Good question? **Yes/No**

Rewritten question: ____________________

____

____

____

____

  ISBN: 9780170425711

b What's the difference between the mean leaf length (cm) of the maple tree and the mean leaf length (cm) of the oak tree in my garden?

Good question? **Yes/No**

Rewritten question: ______________________

c I wonder what the difference is between the arm span of a group of students at Paradise High School?

Good question? **Yes/No**

Rewritten question: ______________________

2 Write appropriate questions to match the following graphs of data from Paradise High School.

a

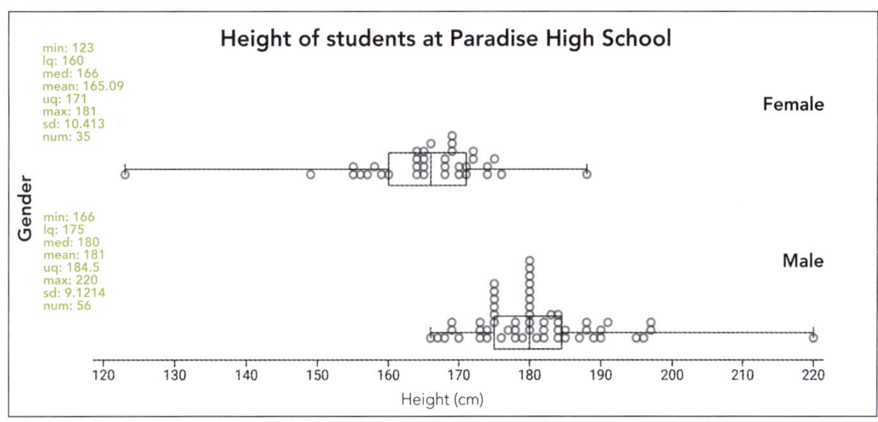

____

b

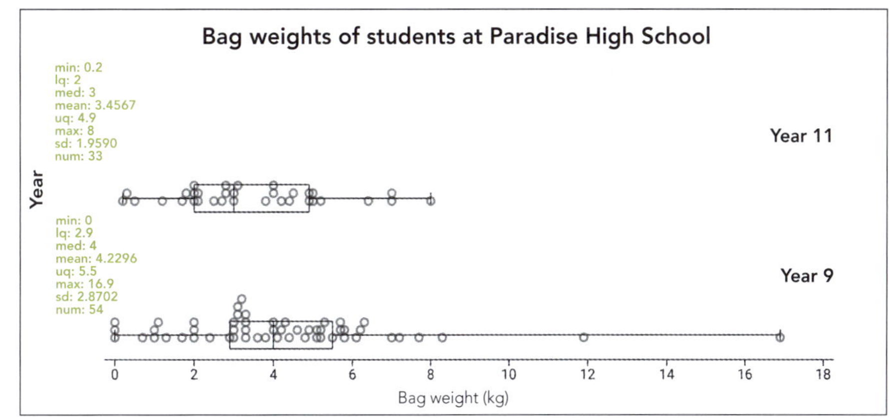

____

ISBN: 9780170425711

3 Define the variables and identify the population

- You must fully define your variables and show that you understand them.
- Further research and explanation as to why the variables were chosen can lead to a more indepth report.
- You must fully define what you consider the population to be.
- This is the population you are making your inference about.

Example: What is the difference between the median **crown length (mm)** of **juvenile** and **adult kea** in **New Zealand**?

You may need to define your subgroups as well as your variables.

Crown length is the distance from the tip of the beak to the back of the skull (occiput) and is measured in millimetres.
The juvenile kea is between 1 and 2.4 years old; an adult kea is over 4 years of age.
Since the kea in our sample came from a variety of locations throughout New Zealand we can consider that the population is all juvenile and adult kea in New Zealand.

4 State the purpose of your investigation and who would find it useful

- Explain **why** you have chosen your variables.
- Discuss **who else** might be interested in your investigation.
- Do **not** discuss what you see in your graphs at this stage.
- Discuss any **prior research** you have done and include **references**.

Example: What is the difference in the median **crown length (mm)** between **juvenile** and **adult kea** in **New Zealand**?

I would like to know if juvenile kea and adult kea have different crown lengths. It seems likely that they will, because most animals grow larger as they grow older. This information could be used to estimate the age of a captured kea.

Frederick J Jannett has shown that weights of skeleton parts, including the skull, can be used to estimate the age of chickens (https://www.jstor.org/stable/2425524?seq=1#page_scan_tab_contents). Apart from this, there appears to be little research into the use of crown and beak lengths for estimating bird age.

ISBN: 9780170425711

Describing and comparing sample distributions

Write comparative sentences to describe the differences and similarities between two groups.

You need to:
- compare the centres
- describe and compare the shape
- compare the positions and sizes of the spreads
- discuss the overlap
- describe any unusual features (there may be none).

Compare the centres

These could be the mean or median.

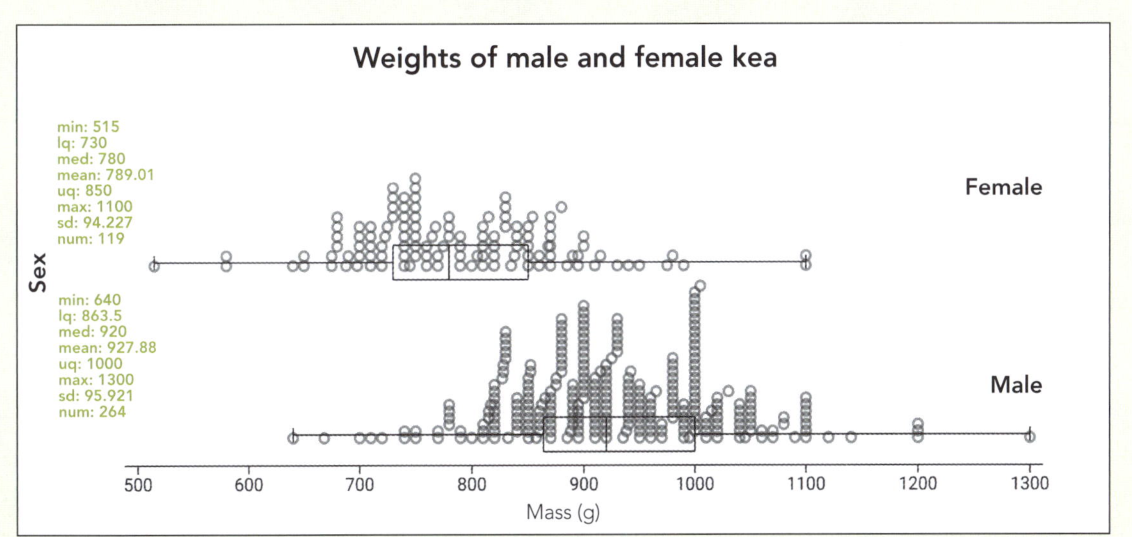

What do you see and what does this mean?

I notice that the median mass for male kea (920 g) is 140 g greater than the median weight of female kea (780 g). This means that, in the sample, male kea were on average heavier than female kea.

ISBN: 9780170425711

Compare the shapes

- It is unlikely that your sample will look exactly like the shapes on page 10. Therefore we use language like 'tends towards …' or 'is approximately …'.

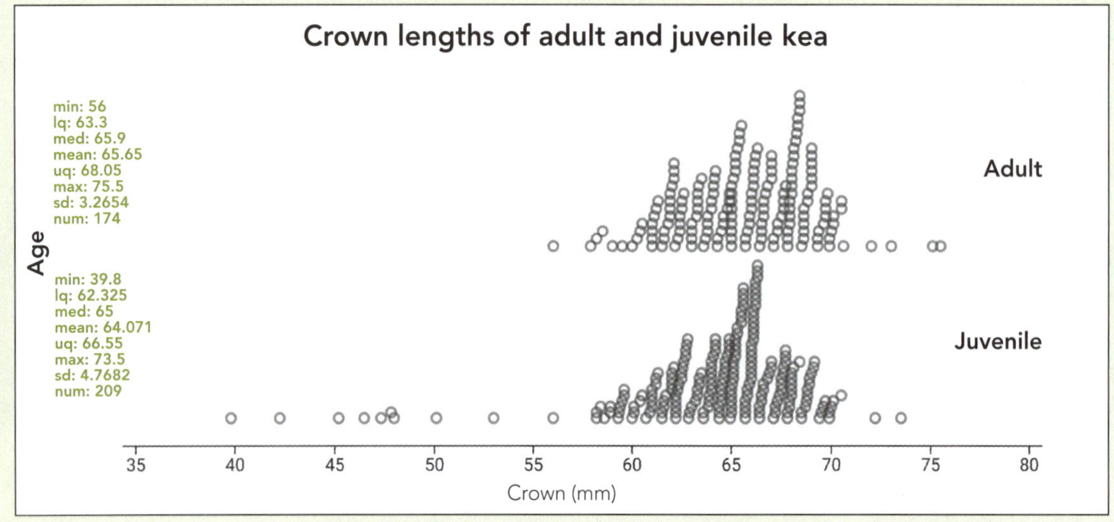

What do you see and what does this mean?

I notice that the crown lengths of adult kea tend towards a bell shaped distribution and the juveniles' crown lengths tend towards a left-skewed distribution. The left skew may be because some of the juveniles haven't fully grown and may be closer to the lower end of the juvenile age range (1 year old) rather than the top of the range at 2.4 years old.

Some kea in the juvenile group may have been less than one year old, and incorrectly categorised as juveniles.

Compare the positions, sizes and overlaps of the spread

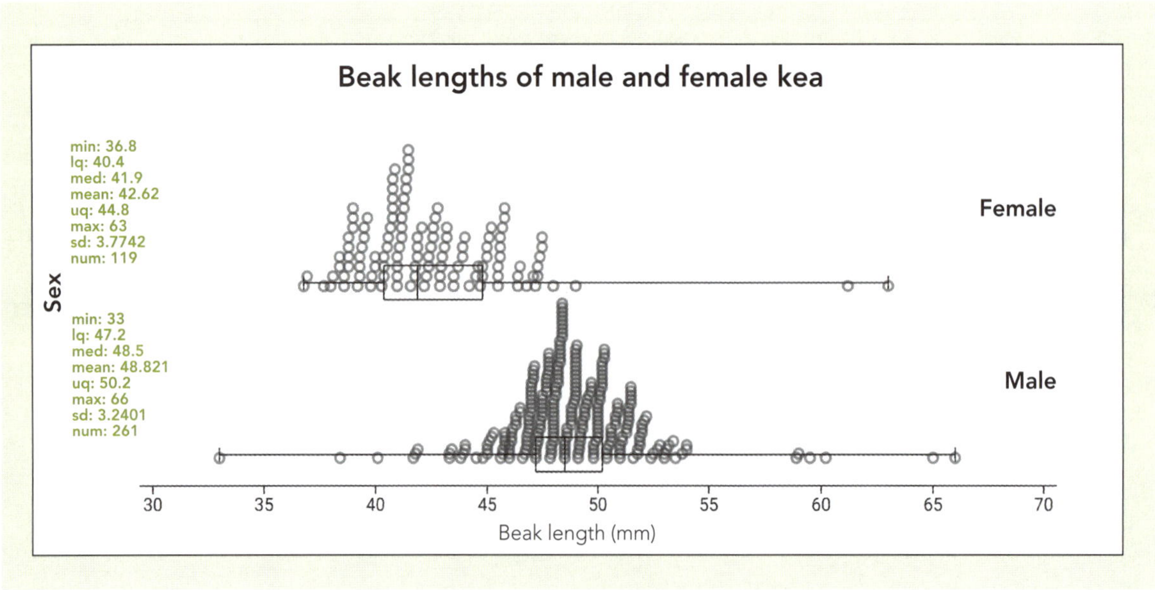

ISBN: 9780170425711

What do you see and what does this mean?

I notice that the interquartile range (IQR) of the beak lengths of male kea (3 mm) is smaller than the IQR for the female kea (4.4 mm). This suggests that the central 50% of female beak lengths is more spread than that for male kea beak lengths.

I also notice that the middle 50% for beak lengths of male kea is further to the right than the female kea, and 75% of the male kea's beak length are greater than 75% of the female kea's beak length. This suggests that in the sample, the male kea have longer beaks than the female kea.

Describe any unusual features

- These could be an **unusual point** or a **cluster**.
- They must be very obvious. Don't comment on them if they are not there!

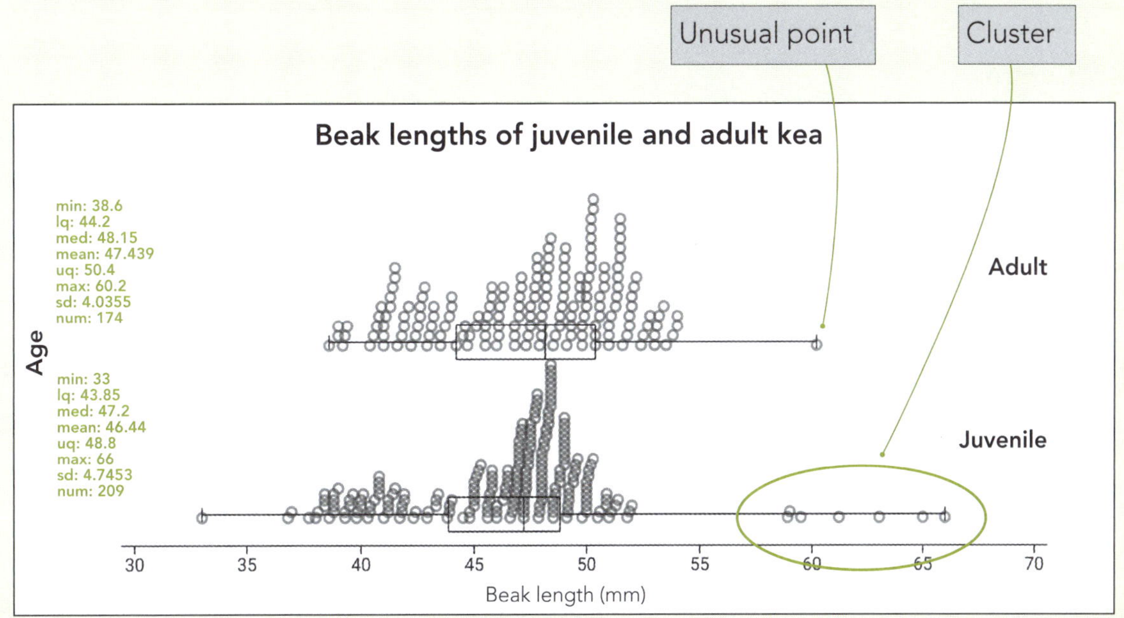

What do you see and what does this mean?

There is one adult kea with an unusually large beak length of 60.2 mm, which is about 6 mm longer than any other adult kea beak length in this sample. This may be because the beak was measured incorrectly or this bird may have an unusually large beak.

There is a cluster of juveniles that also have similar beak lengths, between approximately 58 and 67 mm. These seven juveniles have beak lengths that are at least 5 mm longer than the rest of the juveniles in this sample. This may be because these birds were incorrectly categorised as juveniles. They may be adult kea, although these beak lengths would still be large for an adult.

ISBN: 9780170425711

What do you observe in these graphs?

1 This shows the arm spans of Year 13 students from Paradise High School.

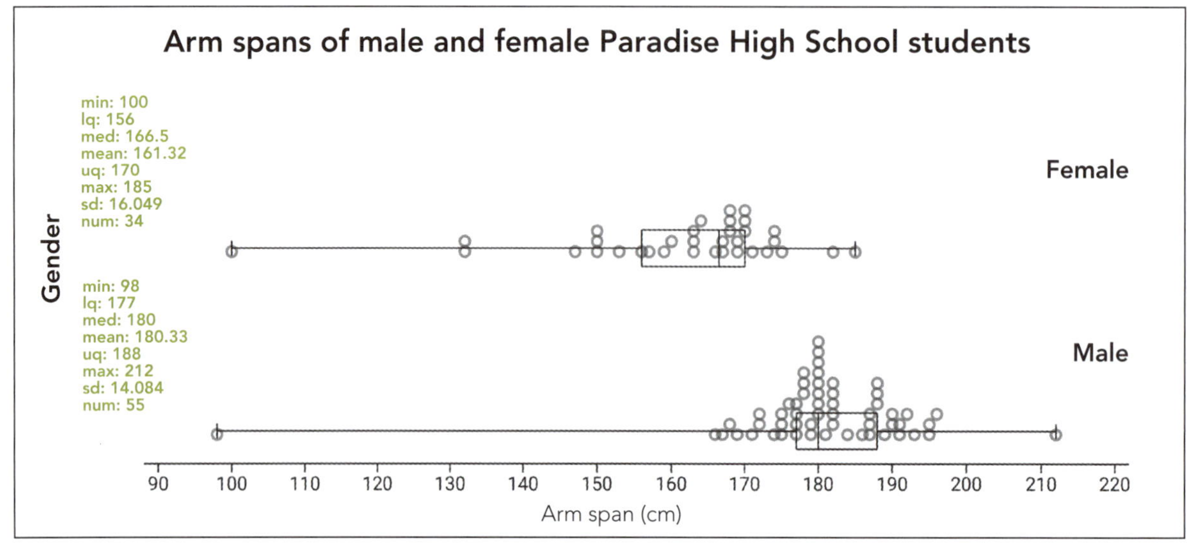

Centre:

____

____

____

____

Shape:

____

____

____

Spread and overlap:

____

____

____

____

____

____

Unusual features:

____

____

____

____

 ISBN: 9780170425711

walkermaths

Data Cards

Create a set of data cards by cutting along the dotted lines.

Kea name	Mass (g)	Beak length (mm)	Crown length (mm)	Age	Sex
Keria	730	41.2	60.4	Juvenile	Female
Marie	770	41.2	57.1	Adult	Female
Patty	760	42.4	61.4	Adult	Female
Scuffle	810	45	61.9	Adult	Female
Anna	730	38.4	58.4	Juvenile	Female
Daisy	580	46.5	61.7	Juvenile	Female
Anthea	750	44	56	Adult	Female
Aotea	805	47.3	62.4	Adult	Female
Mary	800	40.9	62.7	Adult	Female
Swazi	675	38.8	58.2	Juvenile	Female
Pavlova	680	39.3	59.3	Juvenile	Female
Sage	950	42	55	Adult	Female
Trixy	900	42.9	62.7	Adult	Female
Rona	860	46.7	64.2	Adult	Female
Zeal	650	38.8	58.2	Juvenile	Female
Huia	740	39.85	62.75	Juvenile	Female
Zalas	780	45.6	66.6	Adult	Female
Creeky	868	41.5	61.9	Adult	Female
Lake	910	42.4	63.8	Adult	Female
Camel	580	39	56	Juvenile	Female
Rita	775	39.3	61.2	Juvenile	Female
Sparkle	768	41.4	60.9	Juvenile	Female
Cassidy	790	42.5	61	Adult	Female
Goji	750	44.2	65	Adult	Female
Ejay	750	38	48	Juvenile	Female
Zeal	650	38.8	58.2	Juvenile	Female
Medusa	810	38.8	64.2	Juvenile	Female
Xena	830	43.5	67.1	Adult	Female
Rata	835	46	63.6	Adult	Female
Sneaky	840	63	45.2	Juvenile	Female
Hazel	765	41.2	60.1	Juvenile	Female
Flare	725	38.6	60	Juvenile	Female
Beryl	840	41	62.5	Adult	Female
Zelma	830	43.2	63.7	Adult	Female
Templeton	810	38.4	42.25	Juvenile	Female
Handlebars	915	47.5	58.6	Juvenile	Female
Oates	800	41.2	62.6	Juvenile	Female
Gizo	860	44.5	58.2	Adult	Female
Sasha	820	42.5	64.7	Adult	Female

Data Cards

Create a set of data cards by cutting along the dotted lines.

Wagon 970 47.8 69.6 Juvenile Male	Glen 850 49 65.5 Juvenile Male	Ziggy 880 48.4 67.9 Juvenile Male	Sigma 1020 51.7 69.2 Adult Male	Tip Top 110 50.6 69.7 Adult Male
Mazdak 740 47.3 62.2 Juvenile Male	Podrick 1000 48.3 64 Juvenile Male	Aqua 960 44 65 Juvenile Male	Lowrider 1000 53.2 72 Adult Male	Daffy 860 46.8 64.15 Adult Male
Alf 940 47.15 64.95 Juvenile Male	Uri 1020 46.75 67.5 Juvenile Male	Rangi 900 48 68 Juvenile Male	Ranger 940 51.5 65.9 Adult Male	Rimu 980 49.5 67.2 Adult Male
Feisty 780 43.3 64.5 Juvenile Male	Marley 820 41.9 64.6 Juvenile Male	Timbo 1040 49.2 66.1 Juvenile Male	Odyssey 1020 49.3 68.7 Adult Male	Salvatore 1040 40.8 68.3 Adult Male
Tasty 995 48.4 73.5 Juvenile Male	Ocean 1000 49 69 Juvenile Male	Mars 980 48.2 69.9 Juvenile Male	Hoopy 920 46.1 67.3 Adult Male	Cheezel 930 52.2 67.8 Adult Male
Rico 820 46.9 65.7 Juvenile Male	Jasper 920 47.85 65.2 Juvenile Male	Stoney 862 49.2 64.4 Juvenile Male	Fanjo 1300 48.3 66.2 Adult Male	Vick 920 47.6 69.6 Adult Male
Leonardo 950 45.8 65.7 Juvenile Male	Apollo 1000 49 68.8 Juvenile Male	Baggins 820 51.5 68.6 Adult Male	Allsorts 1010 53.5 66.4 Adult Male	Zeus 850 46.2 61.8 Adult Male
Hendrix 865 47 64 Juvenile Male	Benji 940 48 68 Juvenile Male	Herbert 840 48.6 65.2 Adult Male	Bandit 880 46 64.3 Adult Male	The Hulk 920 52.4 65.8 Adult Male

2 This shows the number of Level 2 NCEA credits earned by students from Paradise High School.

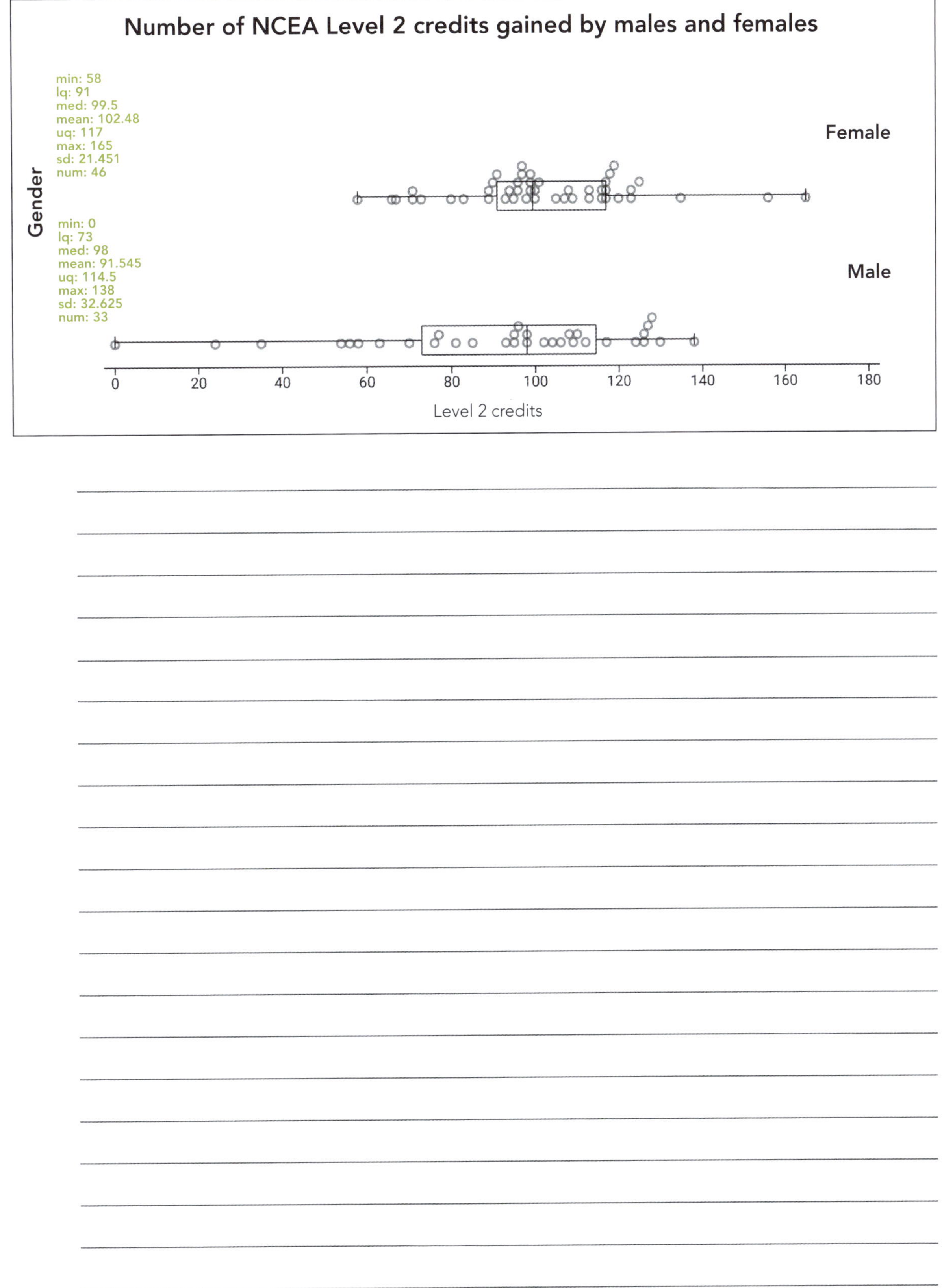

Making a formal inference about the difference between medians

In order to be able to answer your question, you will need to produce a graph similar to the one below.

Example 1: For kea in New Zealand:

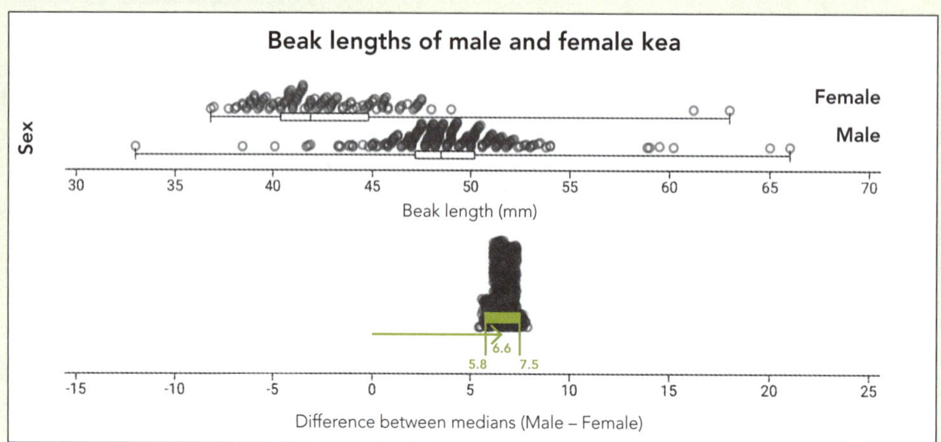

Formal inference using resampling

We can be reasonably confident that for these populations of male and female kea from New Zealand, the ***median*** *beak lengths will be between* ***5.8 mm*** *and* ***7.5 mm longer*** *for male kea than for the female kea.*

Direction, i.e. which group is bigger/larger/longer.

State the confidence intervals in context.

Does your confidence internal contain zero?

As zero ***isn't*** *included in the bootstrapped confidence interval, we* ***can*** *make the call that there is a difference in the median beak lengths between male and female kea from New Zealand.*

Scaffold for when zero is not included in the interval

Population

Variable (i.e. mass)

Groups, e.g. male and female kea

We can be reasonably confident that for the population of ________________ from ________________, the **median** ____________ will be between ____________ and ____________ greater for ____________ than ____________.

Lower limit

Upper limit

First group

Second group

As zero **isn't** included in the bootstrapped confidence interval, we **can** make the call that there is a difference in the **median** ____________ between ____________ and ____________ from ____________.

Variable (i.e. mass)

Groups, e.g. male and female kea

Population

 ISBN: 9780170425711

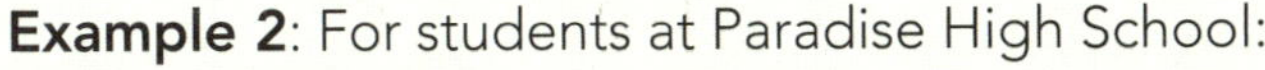

Example 2: For students at Paradise High School:

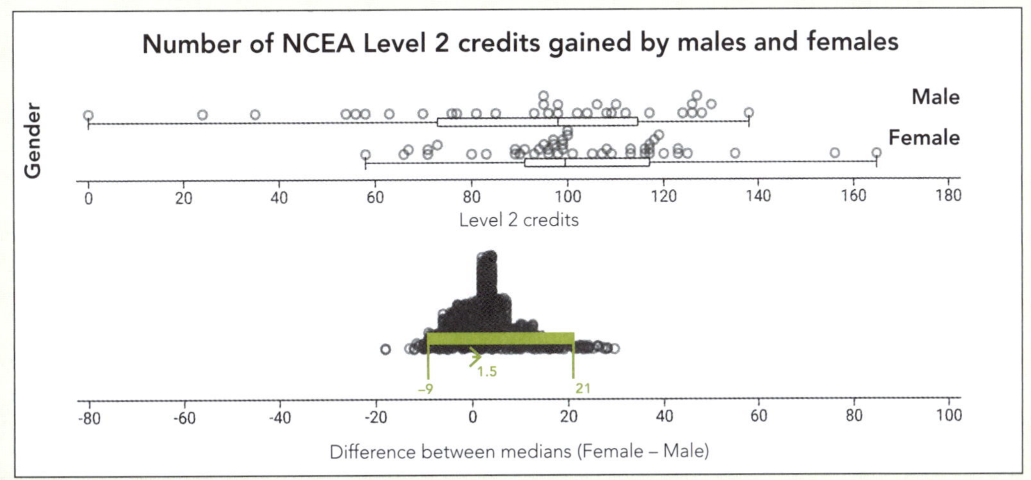

Formal inference using resampling

We can be reasonably confident that for these populations of male and female Level 2 NCEA students from Paradise High School, the median total number of credits earned will be between ***9 fewer*** *and* ***21 credits*** *more for female students than male students.*

This time the confidence interval has 0 within it.

As zero ***is*** *included in the bootstrapped confidence interval, we* ***cannot*** *make the call that there is a difference in the median number of Level 2 credits between males and females from Paradise High School.*

Scaffold for when zero is included in the interval

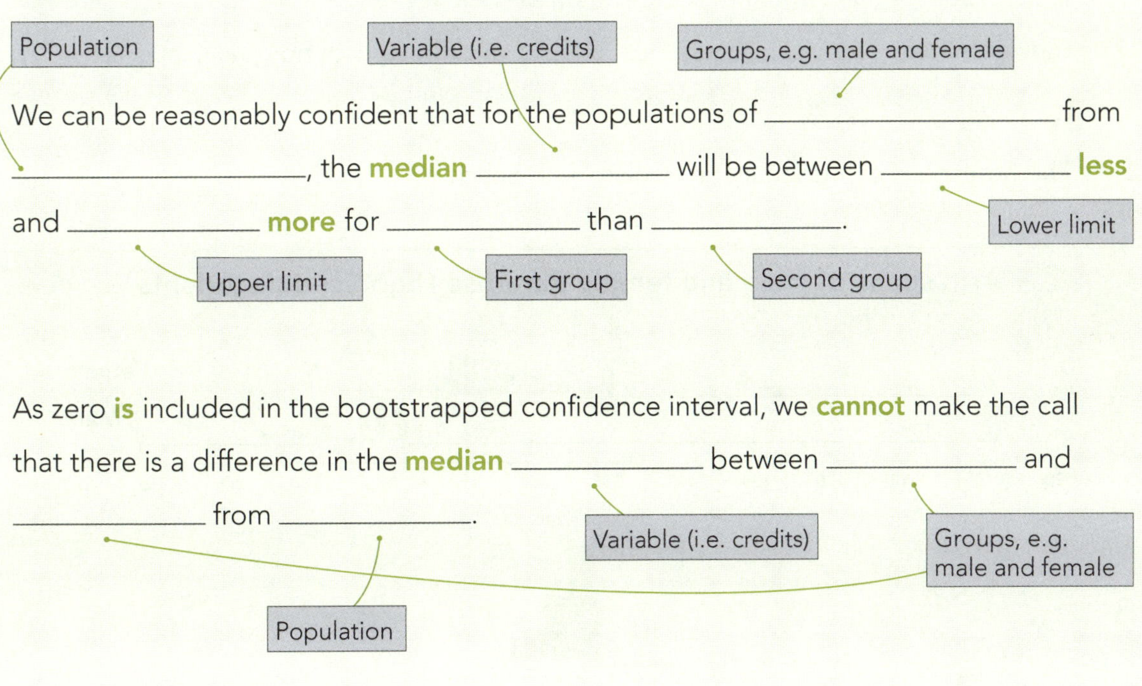

1 This shows the weekly income of New Zealanders who have no formal qualifications and those who have university qualifications.

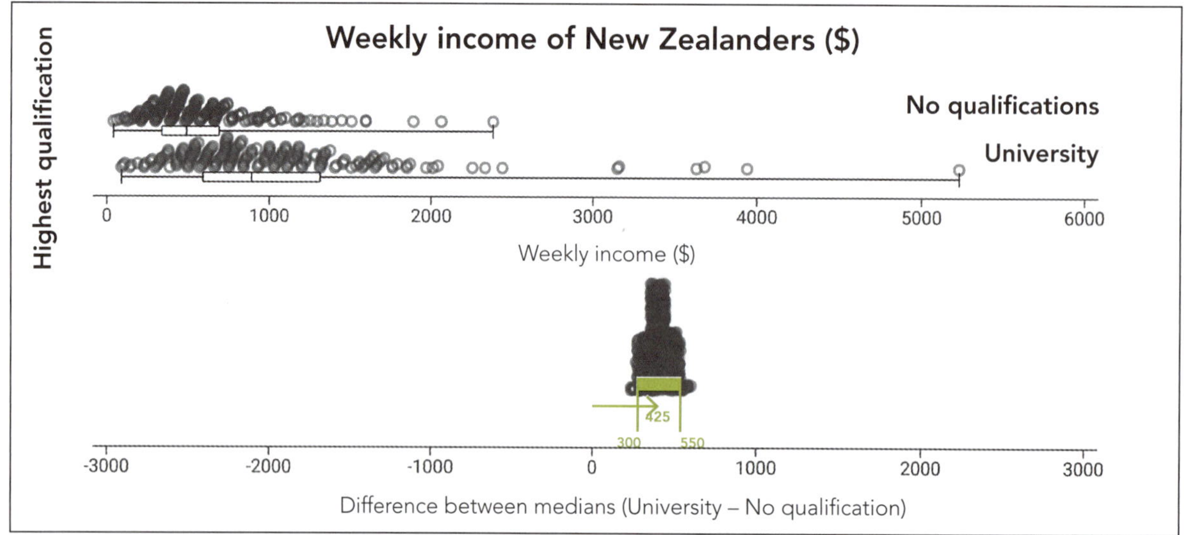

We can be reasonably confident that for the populations of ____________________ ____________________ from ____________________, the median ____________ will be between ____________ and ____________ greater for those with university qualifications than those with ____________________.

As zero ____________ included in the bootstrapped confidence interval, we ____________ conclude that there is a difference in the median ____________ between those ____________________ and those with no qualifications.

2 The arm spans of students at Paradise High School:

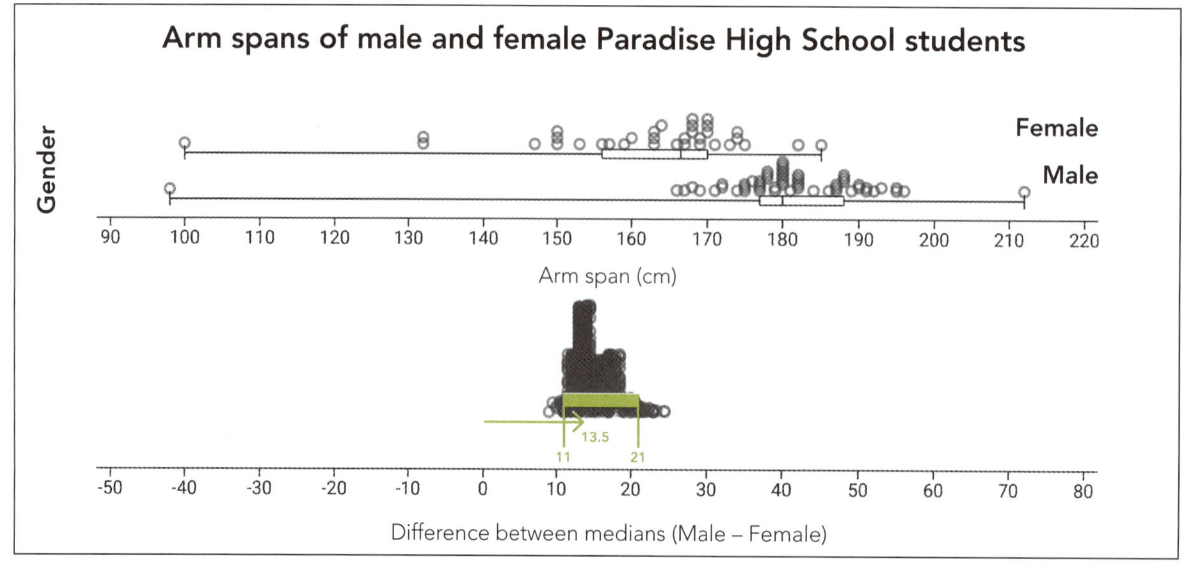

  ISBN: 9780170425711

We can be reasonably confident that ______

3 The heights of Douglas fir and Pinus radiata trees in a forest.

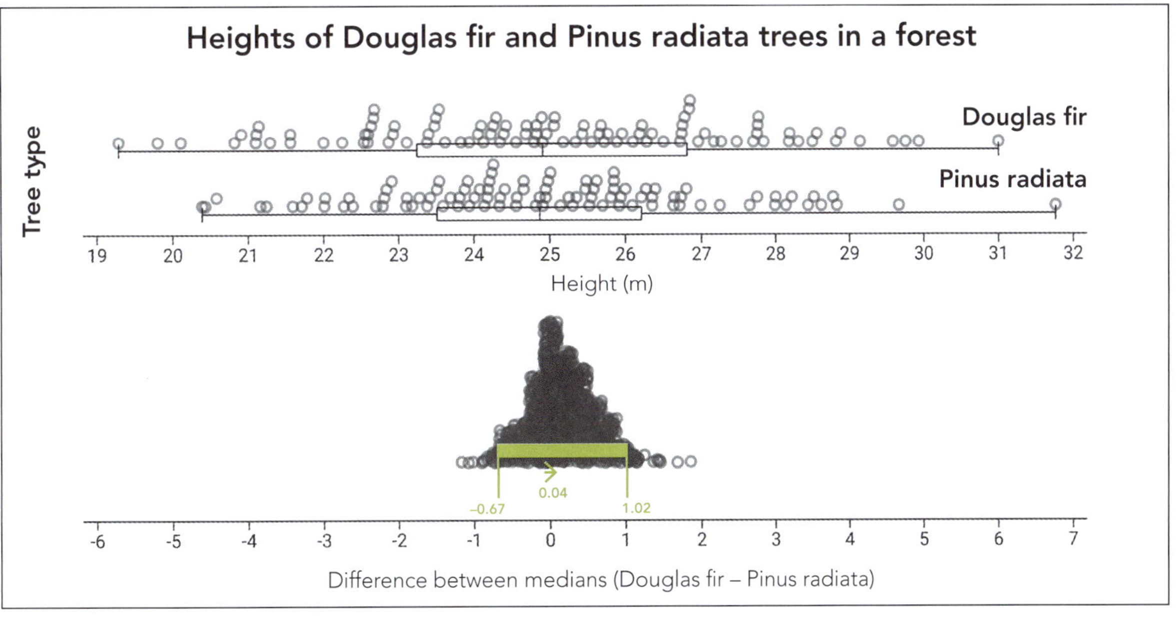

We can be reasonably confident that ______

Making a formal inference about the difference between means

Means should be used only if:

- the data is not skewed
- there are no very unusual points.

Otherwise use the median.

Example:

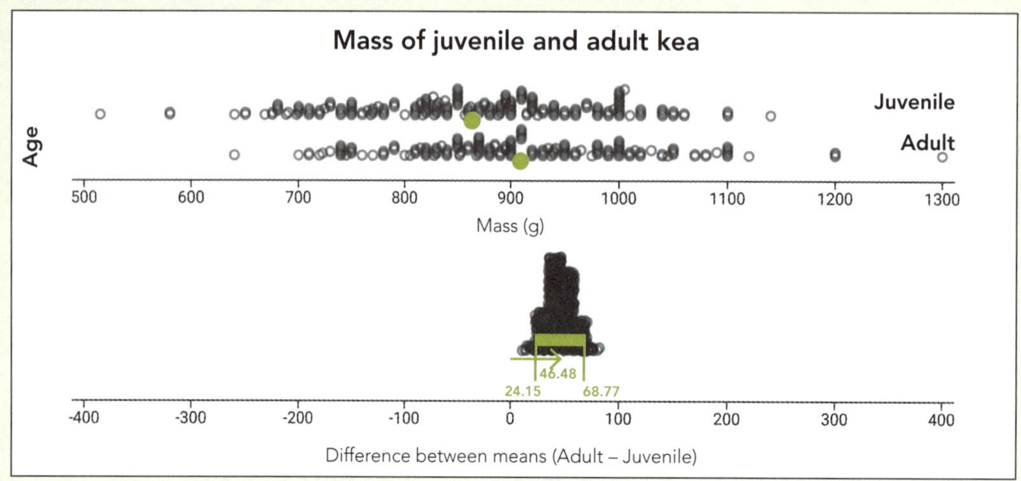

Formal inference using resampling

We can be reasonably confident that for these populations of juvenile and adult kea from New Zealand, the ***mean*** *mass will be between* ***24.15*** *and* ***68.77 grams heavier*** *for adult kea than the juvenile kea.*

As zero ***isn't*** *included in the bootstrapped confidence interval, we* ***can*** *make the call that there is a difference in the mean mass between adult and juvenile kea in New Zealand.*

Scaffold for when zero is not included in the interval

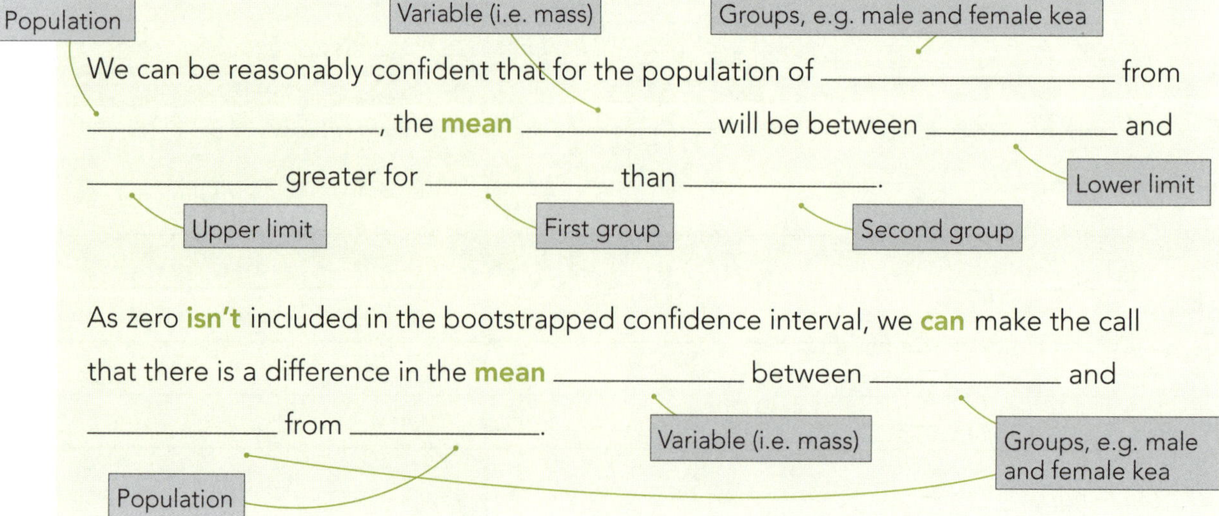

ISBN: 9780170425711

1 The weekly work hours of New Zealanders who have no formal qualifications and those who have university qualifications.

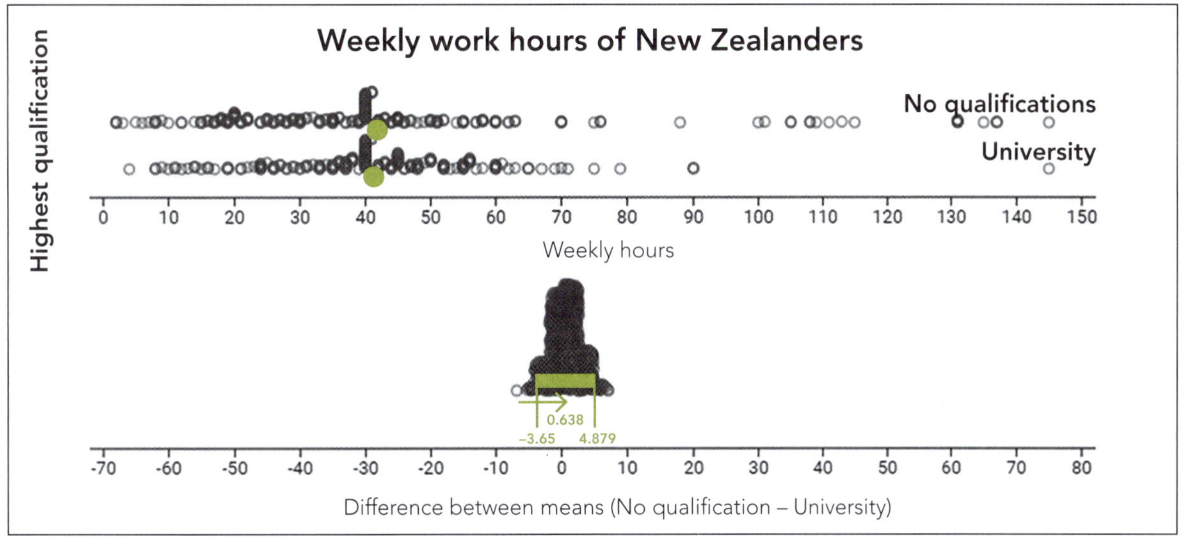

We can be reasonably confident that for the populations of ________________ ________________ from ________________, the mean ________________ will be between ________________ fewer and ________________ more than those with no qualification than those with ________________.

As zero ________________ included in the bootstrapped confidence interval, we ________________ conclude that there ________ a difference in the mean ________________ between those ________________ and those with university qualifications.

2 The girths of Douglas fir and Pinus radiata trees in a forest.

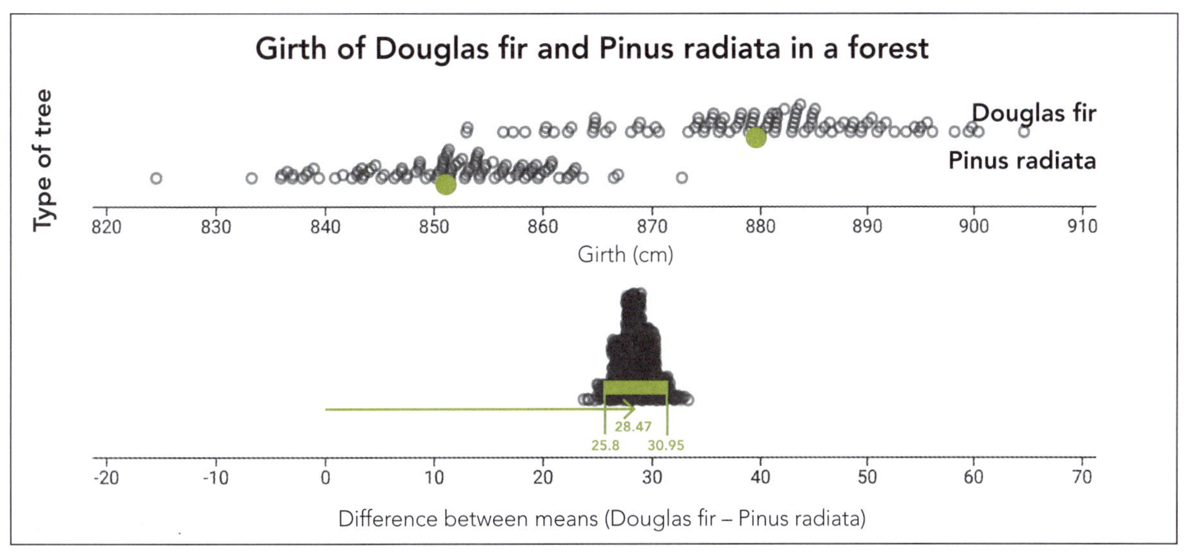

ISBN: 9780170425711

We can be reasonably confident that

3 The bag weights of students from Paradise High School.

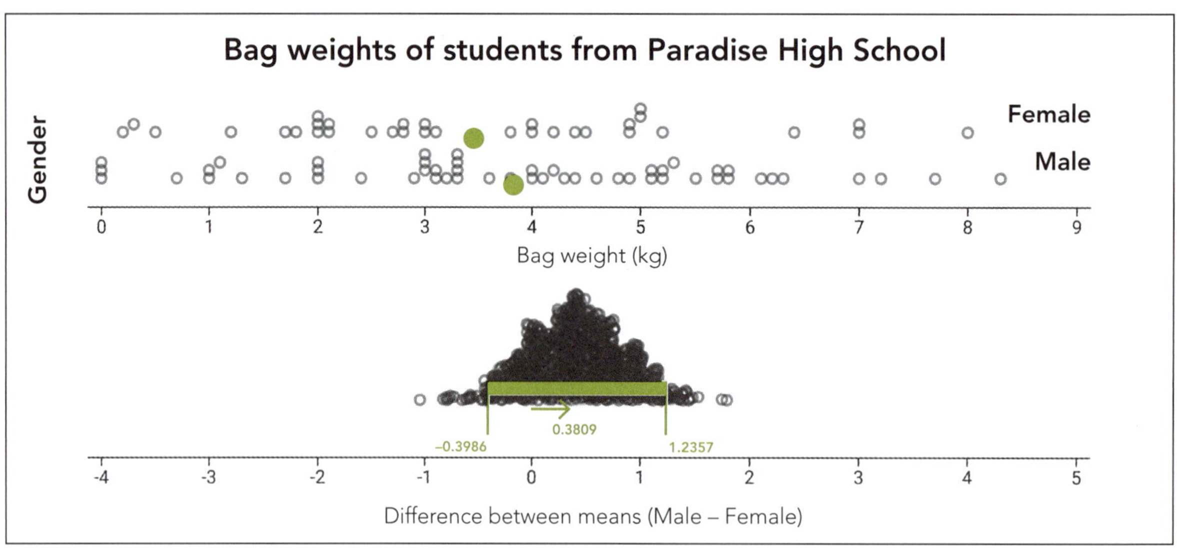

We can be reasonably confident that

ISBN: 9780170425711

Communicating your findings in a conclusion

In your conclusion you should:

- answer your question
- make a call about the differences in the median or mean in the population
- discuss whether the information you have learnt in your report is useful and whether it fulfils your **purpose**
- repeat earlier important points.

Do not include new information! This is where you should be summing up your ideas and coming to your final point.

Writing about statistical inference

When writing about statistical inference, it is important to be clear whether you are writing about the **sample** or the **population**. Here are some examples.

Example:

For the males in the population the data shape is irregular and the females are skewed to the right.

In our sample the male data has an irregular shape and the female data is skewed to the right.

There are very few situations in which you can be definite about what you are stating.

Example:

There is always a difference of at least 6.6 mm in beak length of male and female kea.

The statistics indicate there is an average difference of 6.6 mm in beak length between male and female kea.

In nearly all situations, you should **avoid definite statements**.

Select which is best from each pair of green terms or phrases.

1 I chose to use the mean because **it is appropriate if the data has no unusual points/the data has no unusual points so it is appropriate.**

2 In the graphs we observe that 75% of the amount that male students spent on formal wear is less than 50% of the amount female students spent on formal wear. This **suggests/means** that in the sample, female students on average **do/are likely to** spend a greater sum of money on formal wear than male students.

3 I **could/can** make the call that the median weight for female kiwi was 2.9015 kg and for male kiwis was 2.246 kg, with a difference of 0.6555 kg. This indicates that for this sample, females **are/are likely to be** on average heavier than males.

4 In this sample, 75% of the female athlete body fat percentages are greater than 75% of the male athlete body fat percentages. This **suggests/means** that for this sample, the females have higher body fat percentages on average than males.

5 By looking at the data for weights, researchers might also be able to **suggest/tell** whether or not the kiwi have access to enough food.

6 I notice that in the sample the interquartile range (IQR) of the weights of female kiwi (0.507 kg) is greater than the IQR of male kiwi (0.18 kg). This **suggests/means** that the middle 50% of female kiwi weights are more spread that the middle 50% of male kiwi weights.

7 The spread for the middle 50% of the males is wider than that for the middle 50% of the females. This **means/suggests** that the males have more variation in the sample data than the females.

8 If I were to use the mean as an average for sample distributions that are not symmetrical, it **could/would** affect my investigation because the mean **can/will** be influenced by extreme values.

ISBN: 9780170425711

Pick the errors

The following statements have errors or significant omissions. Identify these.

1 For this report I used the mean to measure the average. I thought this was the best option because the sample was reasonably symmetrical, with the forwards having only 16 more players than the backs.

2 For the populations of both the forwards and the backs, the data shape is irregular.

3 We can be reasonably confident that for the populations of male and female great white sharks from New Zealand, the median total length will be between 5 cm and 9.5 cm.

4 The confidence intervals show that in the sample the median weight for female kiwi will be between 0.597 kg and 0.7115 kg heavier than the median weight for male kiwi.

5 I notice that the IQR for the female group is further to the right (13 cm–21 cm) than the IQR of the male group (7 cm–10 cm).

6 With more data we would expect the right skew shape to become more evident because with more data comes increased sampling variability.

7 The range for the backs is 28 kg and the range for the forwards is 39 kg. This means that the forwards range is 11 kg higher.

8 We can be reasonably confident that, for the New Zealand and South African rugby teams, the forwards will always be between 16 and 23 kg heavier than the backs.

ISBN: 9780170425711

What is the population?

Some sources of sample data are given below. Select the most appropriate population(s) about which you could make an inference in a report.

Sample	Population
1 Babies born in Takapuna Hospital in 1982.	**a** All babies born in New Zealand in 1982. **b** All babies born in 1982 worldwide. **c** Babies born in Takapuna Hospital. **d** All babies born in Takapuna Hospital in 1982.
2 Credit card transactions on Boxing Day in Hamilton.	**a** All credit card transactions from today. **b** All credit card transactions on Boxing Day in New Zealand. **c** All credit card transactions in Hamilton on Boxing Day on the year of the sample. **d** All credit card transactions in Hamilton.
3 The time it takes Paradise High School students to get to school.	**a** The time it takes all Paradise High School students to get to school. **b** The time it takes all students in the world to get to school. **c** The time it takes all teachers at Paradise High School to get to school. **d** The time it takes the student who walk to get to school.
4 Amount of chocolate consumed per person on Christmas Day on Waiheke Island.	**a** The amount of chocolate consumed per person by Waiheke Island residents. **b** The amount of chocolate consumed by each New Zealander on Christmas Day. **c** The amount of chocolate consumed by each person on Waiheke Island on Christmas Day. **d** The amount of chocolate consumed by each North Island resident on Christmas Day.
5 Hours of television watched during the holidays by a sample of Year 11 students from Paradise High School.	**a** Hours of television watched by Year 11 students from Paradise High School all year. **b** Hours of television watched during the holidays by all Year 11 students from Paradise High School. **c** Hours of television watched by all Year 11 New Zealand students in the holidays. **d** Hour of television watched during the holidays by all Paradise High School students.

ISBN: 9780170425711

Putting it all together

The following are the requirements for a basic report. These have been numbered in the report on pages 52–53.

1. Pose an appropriate question informed by contextual knowledge.
2. Identify and fully describe the variables and the population.
3. State the purpose of the investigation and who might find it useful.
4. Produce the appropriate graphs that display sample statistics including a bootstrap interval.
5. Make statements comparing the sample statistics and box plots.
6. Make a formal inference.
7. Give an explanation of sampling variability.
8. Write a conclusion and link it to the purpose of your report.

Optional extras

A Produce indepth research and evidence of contextual knowledge, including sources of information.

B Discuss and justify the use of median or mean bootstrapping.

C Reflect on the process.

D Discuss limitations and make suggestions for further investigation.

ISBN: 9780170425711

Annotated example

Beak lengths of New Zealand kea

What is the difference between the median beak length (mm) of adult kea and the beak length (mm) of juvenile kea in New Zealand? The sample used for this report was provided by the Department of Conservation.

1 Appropriate question

Those who might find this report useful would be researchers who are interested in collecting anatomical data on kea. This could help scientists to predict bird age. This may be useful for aging captured birds for which there is no record. This may be of particular use to the Kea Conservation Trust, who design kea playgrounds and monitor kea sightings around New Zealand (http://www.keaconservation.co.nz/).

3 Purpose.

A Research and sources of information

The beak length has been measured in millimetres. This picture shows how the beak length is measured.

2 Variables and population identified

A juvenile kea is one that is considered to be between 1 and 2.4 years old. An adult kea is aged 4 years or older.

Since the kea in our sample come from a variety of locations throughout New Zealand, we can consider that the population is all juvenile and adult kea in New Zealand.

4 Data displayed

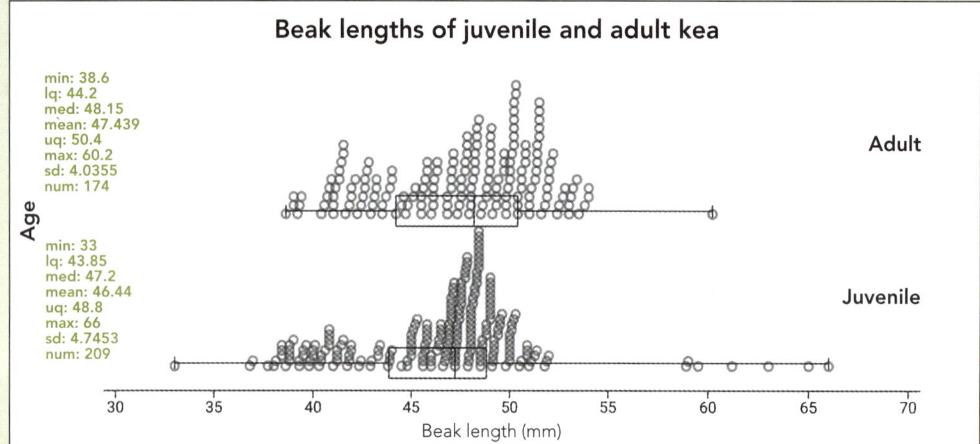

In the samples the adult median beak length is 0.95 mm longer than that of juveniles. The middle 50% of the adult kea data is slightly more spread out and further to the right than the middle 50% of the data for juveniles.

Most of the middle 50% of juvenile kea beak length data overlaps with the middle 50% of adult kea beak length data for these samples.

The juvenile keas' IQR is 4.95 mm whereas the adult keas' IQR is 6.2 mm. So even though the range is larger in the juvenile sample the middle 50% of juvenile kea beak lengths have less variability than the adult kea beak lengths.

5 Statement comparing sample statistics and box plots

The shape of both sets tends towards bimodal, with the larger mode on the right.

There are unusual features in both groups. There is an adult kea with a beak length of 60.2 mm, which is 6.2 mm larger than the closest beak length. In the juveniles' group, there is a cluster between 58 and 67 mm. This is unusual and it may suggest that those birds have been aged incorrectly and are actually adult kea. However, these beak sizes would still be large for an adult kea.

ISBN: 9780170425711

Overall, in the samples, if the unusual features were removed the distributions would be similar with adult kea having slightly longer beaks than juveniles.

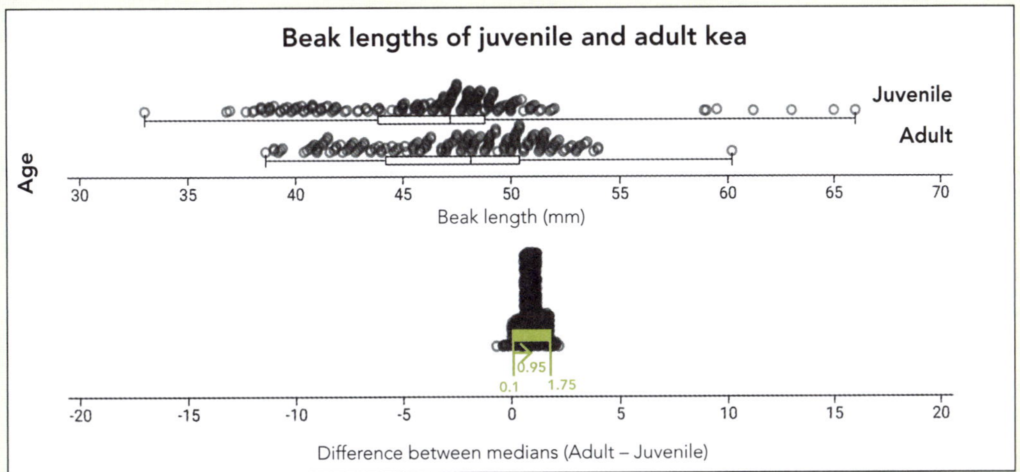

4 Bootstrap interval displayed

We can be reasonably confident that for these populations of adult and juvenile kea from New Zealand, the median beak lengths will be between 0.1 and 1.75 mm longer for adult kea than for juvenile kea.

As zero isn't included in the bootstrapped confidence interval, we can make the call that there is a difference in the median beak lengths between adult and juvenile kea from New Zealand.

6 Make a formal inference

I am assuming that my original sample of 174 adult kea and 207 juvenile kea are representative of the population of all kea. If I were to take another sample, the displays and sample statistics might be different as the sample would probably contain different kea.

7 Sampling variability explained

Based on these samples, I can make the call that there is a difference in the beak lengths between New Zealand adult kea and New Zealand juvenile kea. I can make the call that the adults are likely to have a longer median beak length based on the bootstrapped confidence interval.

For this report, I used the median as a measure of average. I felt that this was appropriate as both sample distributions were fairly symmetrical. Because they were symmetrical, I could have also used the mean. This is evident in the sample statistics where the difference between the sample means (0.99 mm) is similar to the difference in the sample medians (0.95 mm).

B Use of the mean or median

The bootstrapped interval suggests that there is a small difference between the beak lengths of adult and juvenile kea. However, because there is a lot of overlap in the sample data and the sample medians differ by less than 1 mm, this difference in the population is likely to be small.

C Reflection on process

Because most of the middle 50% for these two samples overlap, and there is a minimum possible beak length difference of 0.1 mm (lower limit of the confidence interval) between adults and juveniles in the population, using beak length to predict age is likely to be imprecise. Therefore scientists would be unwise to rely solely on beak length to predict kea age.

8 Conclusion linked to purpose

It is likely that a number of different people, including volunteers, took these measurements and perhaps not all measured beak lengths in the same way. I would also be sceptical that a kea would stay still long enough for someone to get an accurate measurement.

D Limitations and further investigation

ISBN: 9780170425711

Practice tasks

Practice task one

Find the eight major mistakes or omissions from this report. Identify each and write a corrected version for each.

Pizza delivery times
Is the median delivery time for Pizzarella longer than the median delivery time for Crusty?

The variables I will be using are the times, in minutes, taken by each company to make a pizza. That way I will know which store takes longer to deliver pizzas.

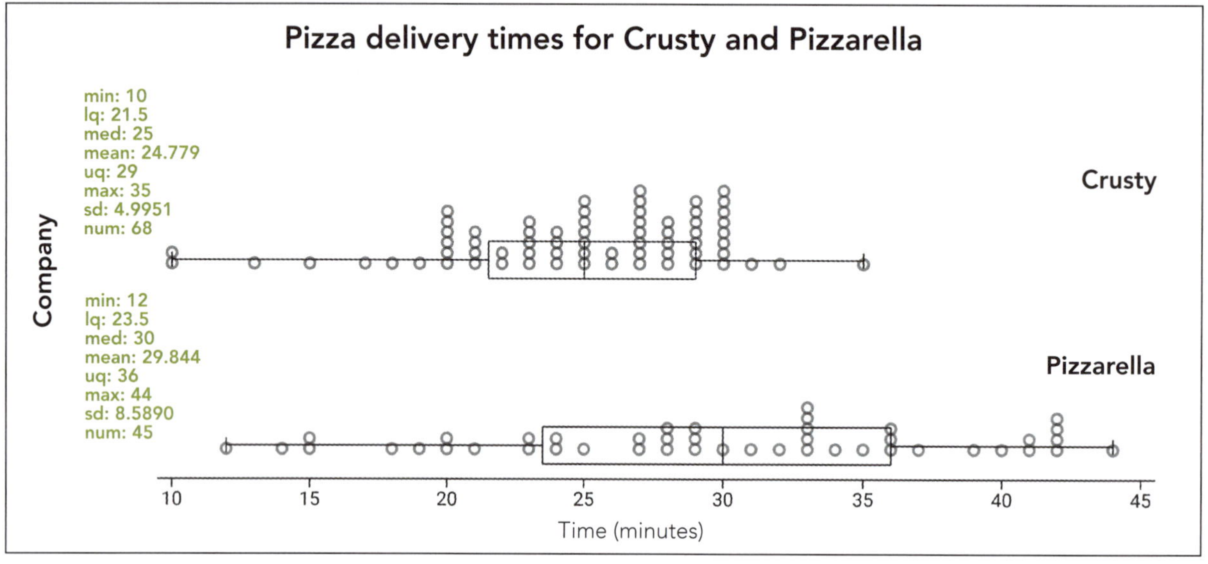

Crusty's median delivery time is 24.779 minutes, whereas Pizzarella's median delivery time is 30 minutes. The middle 50% of Pizzarella delivery times are further to the right (i.e. they take longer) than the Crusty company.

The median delivery time of Pizzarella is higher than the upper quartile for Crusty. This means that 75% of Crusty's deliveries take longer than half of Pizzarella's.

The shape of the distribution times for both companies tends towards irregular although, because the interquartile range of Crusty is smaller, they are more consistent with their delivery times.

 ISBN: 9780170425711

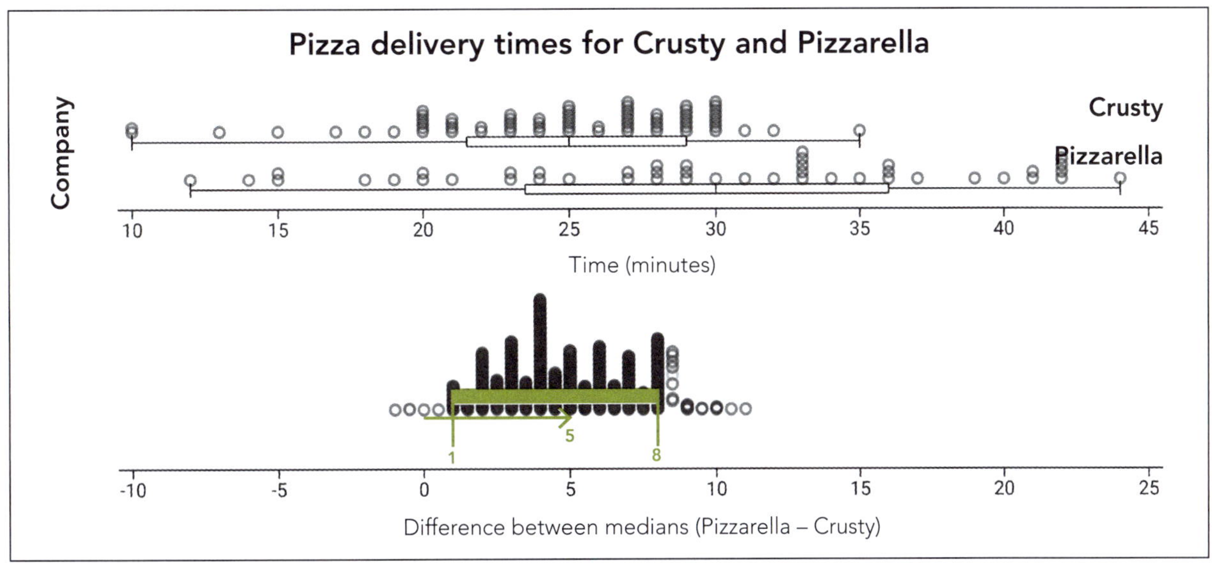

I am confident that for these pizza delivery companies, the difference between the median delivery time is between 5 and 8 minutes.

This means that Crusty's pizza delivery is always going to be faster than Pizzarella's.

If I did this project again, I would probably get different displays and sample statistics because the companies might have changed owners.

1 ____

2 ____

ISBN: 9780170425711

3

4

5

6

7

8

 ISBN: 9780170425711

Practice task two

Sixty-two hockey players were wearing tracking devices for their weekend game. The distance they ran (km) during the game was recorded. At the end of the game, this data was collated and the hockey players' positions were noted. The results are recorded in kilometres.

An electronic version of the data can be found at www.cengage.co.nz/3.10-statistical-inference.

This activity requires you to carry out a statistical inference investigation using the given variables and to write a report of your findings.

Use the statistical enquiry cycle to carry out a statistical investigation to determine if there is a difference in the medians between the given variables.

Write a report describing the investigation.

1 Familiarise yourself with the data set provided. This will include doing research to help you understand the variables and develop a purpose for the investigation.

2 Identify the variables you wish to investigate, and establish a related investigative comparison question.

3 Conduct your investigation and write a report containing:
   - your comparison investigative question
   - appropriate displays and summary statistics
   - a discussion of the sample distributions
   - an appropriate formal statistical inference
   - a conclusion communicating your findings, including discussing sampling variability, the variability of estimates, and reflecting on the process that has been used to make the formal inference.

In writing your report, link your discussion to the context and support the statements you make by referring to statistical evidence.

Position	Distance covered (km)
Offence	9.92
Offence	9.62
Offence	6.81
Offence	8.61
Offence	6.43
Offence	8.53
Offence	9.43
Offence	8.39
Offence	9.19
Offence	10.21
Offence	9.16
Offence	8.67
Offence	6.84
Offence	9.76
Offence	8.69
Offence	8.81
Offence	9.01
Offence	8.9
Offence	9.54
Offence	9.74
Offence	9.14
Offence	8.48
Offence	8.06
Offence	8.83
Offence	9.4
Offence	9.83
Offence	9.87
Offence	9.72
Offence	8.81
Offence	9.55
Offence	8.27

Position	Distance covered (km)
Defence	4.93
Defence	5.74
Defence	8.15
Defence	5.76
Defence	5.94
Defence	7.97
Defence	5.17
Defence	6.05
Defence	6.24
Defence	8.58
Defence	4.57
Defence	6.72
Defence	6.34
Defence	5.5
Defence	6.12
Defence	5.22
Defence	5.95
Defence	6.93
Defence	7.34
Defence	4.91
Defence	5
Defence	5.09
Defence	6.26
Defence	4.51
Defence	5.1
Defence	5.76
Defence	8.8
Defence	6.57
Defence	5.37
Defence	7
Defence	0.6

Question and purpose:

____

____

____

____

Define your variables and population:

____

____

____

ISBN: 9780170425711

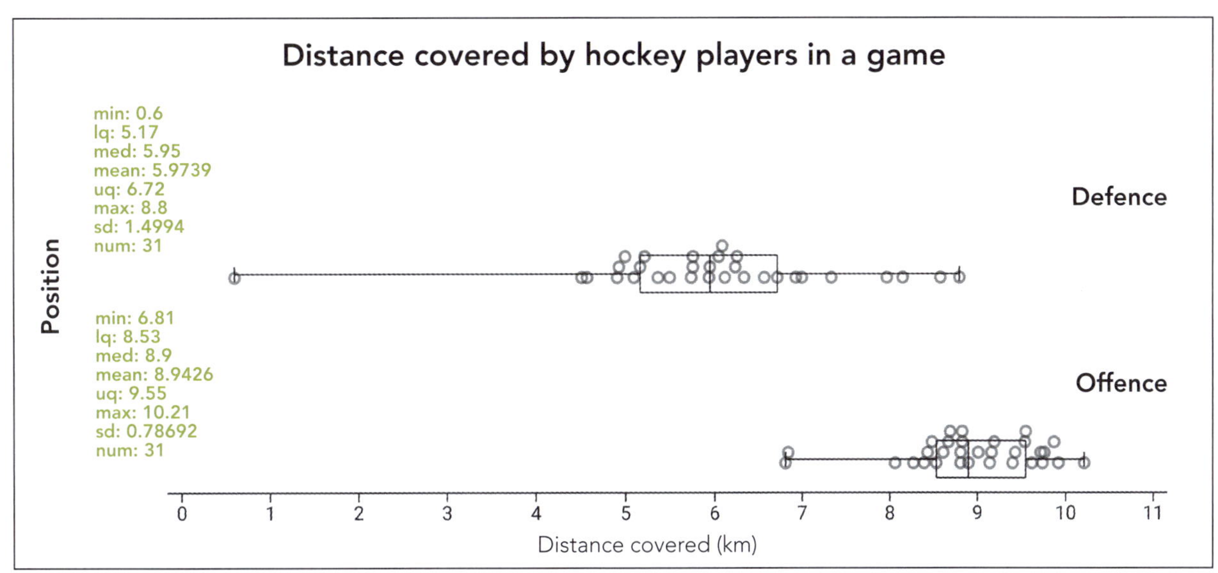

Discuss and compare sample distributions:

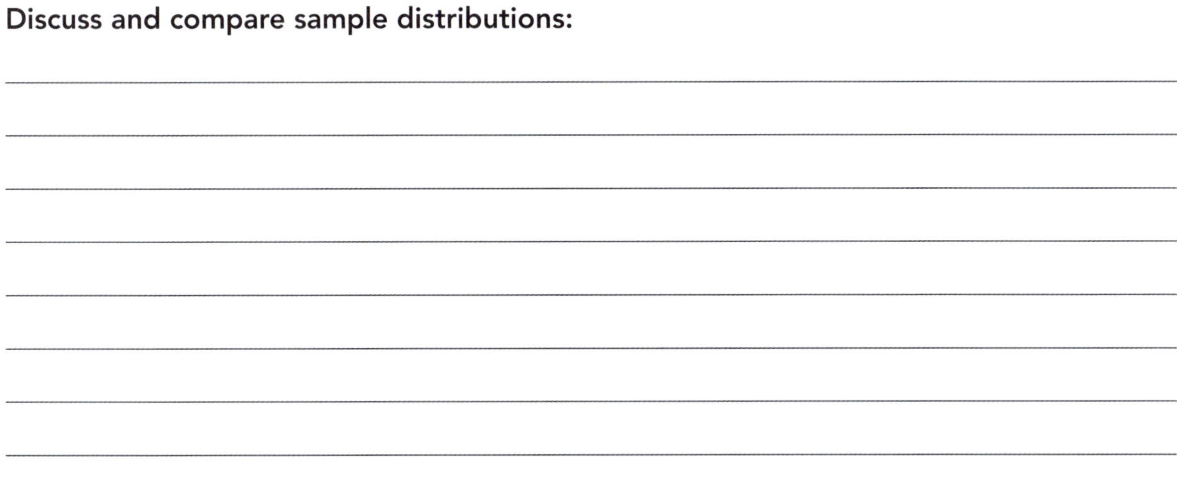

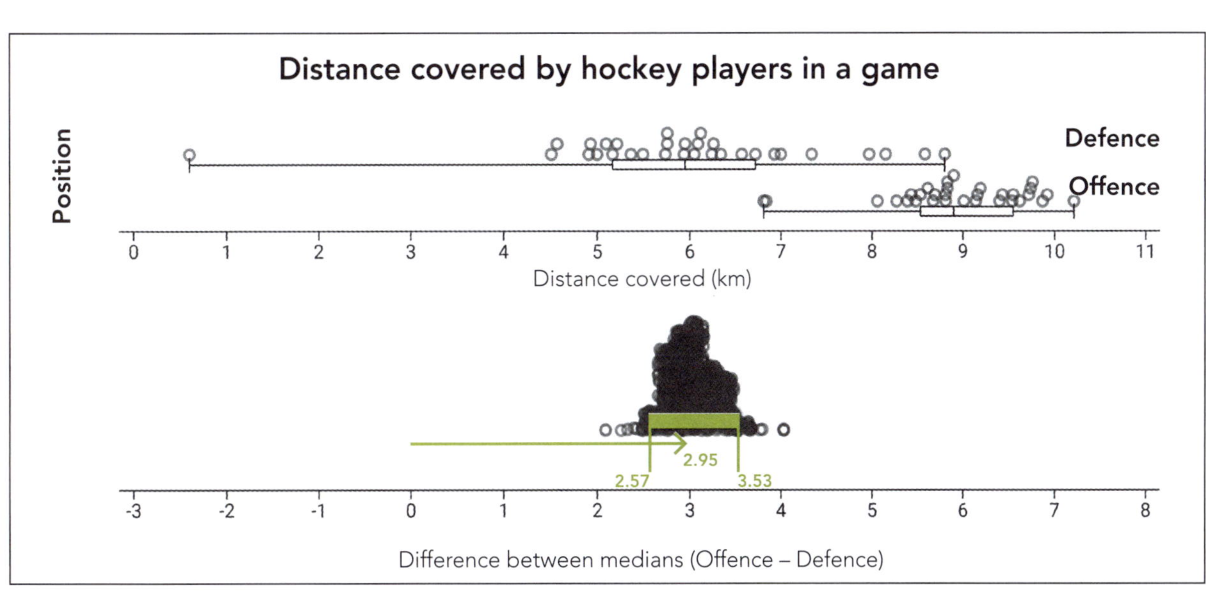

ISBN: 9780170425711

Make a formal inference:

Discuss sampling variability:

Conclusion:

Reflecting, limitations and further investigation:

 ISBN: 9780170425711

Practice task three

A number of participants in the New York Marathon compete in a category for athletes with a disability. This allows them to compete in either a wheelchair or a handcycle alongside the able-bodied competitors. The website https://www.tcsnycmarathon.org/ lists these competitors from the 2016 and 2017 New York marathons. You have been provided with a sample of these competitors.

An electronic version of the data can be found at www.cengage.co.nz/3.10-statistical-inference.

This activity requires you to carry out a statistical inference investigation using the given variables and to write a report of your findings.

Use the statistical enquiry cycle to carry out a statistical investigation to determine if there is a difference in the medians between the given variables.

Write a report describing the investigation.

1 Familiarise yourself with the data set provided. This will include doing research to help you understand the variables and develop a purpose for the investigation.
2 Identify the variables you wish to investigate, and establish a related investigative comparison question.
3 Conduct your investigation and write a report containing:
   - your comparison investigative question
   - appropriate displays and summary statistics
   - a discussion of the sample distributions
   - an appropriate formal statistical inference
   - a conclusion communicating your findings, including discussing sampling variability, the variability of estimates, and reflecting on the process that has been used to make the formal inference.

In writing your report, link your discussion to the context and support the statements you make by referring to statistical evidence.

The data set provides the following variables:

Age	Age
Gender	Male or female
IAAF	International Association of Athletics Federations
Half	The time (minutes) at the halfway mark
Official	The time (minutes) at the finish line
Discipline	Wheelchair or Handcycle
Year	Year in which completed the marathon

Age	Gender	IAAF	Half (min)	Overall (min)	Discipline	Year
49	M	FRA	43.32	88.80	Handcycle	2016
48	M	GER	44.08	88.95	Handcycle	2016
38	M	BRA	43.35	89.25	Handcycle	2016
45	M	ESP	43.97	93.10	Handcycle	2016
50	M	USA	46.20	94.32	Handcycle	2016
34	M	USA	46.20	94.75	Handcycle	2016
44	M	BRA	48.23	97.17	Handcycle	2016
48	M	FRA	48.28	99.98	Handcycle	2016
59	M	USA	48.70	101.38	Handcycle	2016
46	M	USA	48.70	101.63	Handcycle	2016
47	M	USA	50.48	104.20	Handcycle	2016
69	M	USA	51.22	105.18	Handcycle	2016
29	M	USA	51.53	105.70	Handcycle	2016
61	M	USA	51.55	107.47	Handcycle	2016
50	M	USA	54.97	114.05	Handcycle	2016
48	F	NZL	55.65	114.15	Handcycle	2016
47	M	USA	54.65	115.42	Handcycle	2016
35	M	USA	59.27	123.38	Handcycle	2016
42	M	ITA	61.32	128.72	Handcycle	2016
70	M	USA	62.23	131.73	Handcycle	2016
44	M	USA	63.03	134.62	Handcycle	2016
34	M	USA	67.60	139.20	Handcycle	2016
38	F	USA	68.62	140.98	Handcycle	2016
55	M	NZL	71.60	142.78	Handcycle	2016
32	M	USA	71.72	149.77	Handcycle	2016
68	M	USA	77.43	157.05	Handcycle	2016
63	F	USA	77.85	163.03	Handcycle	2016
33	M	USA	78.12	163.43	Handcycle	2016
32	M	USA	82.08	174.70	Handcycle	2016
27	F	USA	82.03	178.60	Handcycle	2016
34	F	USA	86.40	184.40	Handcycle	2016
43	M	GER	91.65	187.10	Handcycle	2016
39	M	USA	110.62	211.27	Handcycle	2016
41	F	USA	95.83	214.13	Handcycle	2016
48	F	USA	98.32	218.28	Handcycle	2016
29	M	USA	114.55	239.95	Handcycle	2016
87	F	USA	132.62	258.13	Handcycle	2016
51	F	USA	123.73	269.32	Handcycle	2016
30	F	VEN	129.78	301.25	Handcycle	2016
63	F	USA	103.33	374.88	Handcycle	2016
35	M	RSA	183.45	398.18	Handcycle	2016
59	M	USA	329.72	571.23	Handcycle	2016
31	M	SUI	46.52	97.35	Wheelchair	2016
27	M	GBR	48.95	99.67	Wheelchair	2016
25	M	JPN	49.05	99.85	Wheelchair	2016
44	M	RSA	49.02	99.93	Wheelchair	2016
46	M	FRA	49.02	99.97	Wheelchair	2016
38	M	ESP	49.10	100.02	Wheelchair	2016

ISBN: 9780170425711

Age	Gender	IAAF	Half (min)	Overall (min)	Discipline	Year
32	M	CAN	46.53	101.13	Wheelchair	2016
43	M	JPN	49.07	102.45	Wheelchair	2016
28	M	USA	49.28	104.30	Wheelchair	2016
32	F	SUI	52.55	108.15	Wheelchair	2016
25	M	CHN	49.08	110.85	Wheelchair	2016
28	F	USA	52.78	111.03	Wheelchair	2016
42	M	USA	55.82	114.87	Wheelchair	2016
22	M	USA	58.15	118.60	Wheelchair	2016
33	F	CHN	58.32	123.33	Wheelchair	2016
23	F	AUS	59.75	124.47	Wheelchair	2016
32	F	GER	60.33	127.38	Wheelchair	2016
20	F	USA	61.10	127.45	Wheelchair	2016
21	F	GBR	61.12	128.55	Wheelchair	2016
26	F	USA	64.70	131.72	Wheelchair	2016
24	F	USA	64.82	132.48	Wheelchair	2016
23	M	USA	66.72	132.62	Wheelchair	2016
20	F	USA	68.57	140.08	Wheelchair	2016
25	M	GBR	66.27	142.52	Wheelchair	2016
46	F	CAN	68.75	143.37	Wheelchair	2016
37	M	NZL	72.35	143.77	Wheelchair	2016
35	F	ESP	76.90	158.37	Wheelchair	2016
41	M	CHI	78.55	161.12	Wheelchair	2016
53	M	USA	76.67	167.05	Wheelchair	2016
33	F	VEN	91.12	196.92	Wheelchair	2016
20	M	NZL	106.98	222.32	Wheelchair	2016
45	F	FRA	115.67	249.18	Wheelchair	2016
30	F	NZL	125.43	255.95	Wheelchair	2016
40	F	VEN	115.40	265.13	Wheelchair	2016
51	M	USA	177.07	413.63	Wheelchair	2016
59	M	USA	208.60	446.88	Wheelchair	2016
65	M	USA	211.27	468.48	Wheelchair	2016
48	M	AUS	40.20	83.10	Handcycle	2017
44	M	ESP	44.53	91.25	Handcycle	2017
26	M	COL	45.72	95.57	Handcycle	2017
36	M	COL	46.93	95.62	Handcycle	2017
43	M	BRA	47.42	101.00	Handcycle	2017
49	M	USA	50.65	104.08	Handcycle	2017
48	M	USA	50.62	105.78	Handcycle	2017
46	M	USA	52.42	110.78	Handcycle	2017
49	M	USA	53.53	111.13	Handcycle	2017
32	M	USA	58.18	117.48	Handcycle	2017
26	M	USA	55.10	118.00	Handcycle	2017
46	M	USA	55.98	121.02	Handcycle	2017
47	M	USA	61.48	125.88	Handcycle	2017
46	M	USA	58.32	126.92	Handcycle	2017
53	M	USA	62.90	130.53	Handcycle	2017
41	M	USA	64.82	134.87	Handcycle	2017
43	M	USA	68.87	143.53	Handcycle	2017

ISBN: 9780170425711

Age	Gender	IAAF	Half (min)	Overall (min)	Discipline	Year
68	F	USA	69.10	144.10	Handcycle	2017
57	M	USA	74.37	149.57	Handcycle	2017
33	M	MGL	62.42	150.75	Handcycle	2017
34	M	USA	73.88	152.68	Handcycle	2017
44	M	USA	81.63	170.32	Handcycle	2017
30	F	USA	80.93	176.80	Handcycle	2017
50	M	USA	86.07	179.12	Handcycle	2017
71	M	USA	82.48	181.75	Handcycle	2017
37	F	USA	83.95	186.88	Handcycle	2017
30	F	USA	88.50	201.18	Handcycle	2017
40	F	USA	94.32	206.10	Handcycle	2017
26	F	USA	96.50	209.50	Handcycle	2017
55	M	USA	100.82	210.02	Handcycle	2017
75	M	USA	105.62	227.03	Handcycle	2017
28	M	USA	117.17	244.15	Handcycle	2017
86	F	USA	126.32	261.70	Handcycle	2017
56	M	USA	133.55	303.57	Handcycle	2017
80	M	USA	133.68	303.82	Handcycle	2017
61	F	USA	144.83	319.00	Handcycle	2017
36	M	USA	149.65	329.80	Handcycle	2017
58	M	USA	179.50	391.43	Handcycle	2017
50	F	USA	217.97	411.73	Handcycle	2017
30	M	SUI	45.68	95.82	Wheelchair	2017
43	M	RSA	47.97	100.13	Wheelchair	2017
34	M	GBR	49.00	100.20	Wheelchair	2017
30	M	USA	47.98	103.67	Wheelchair	2017
29	M	USA	50.78	104.45	Wheelchair	2017
37	M	ESP	51.47	106.73	Wheelchair	2017
27	F	USA	52.47	107.72	Wheelchair	2017
46	M	JPN	51.50	107.88	Wheelchair	2017
31	F	SUI	52.58	109.47	Wheelchair	2017
30	F	USA	52.82	113.25	Wheelchair	2017
39	M	MEX	53.60	114.97	Wheelchair	2017
25	F	USA	52.83	118.27	Wheelchair	2017
26	M	USA	56.40	119.60	Wheelchair	2017
19	F	USA	58.75	123.03	Wheelchair	2017
20	F	GBR	58.67	123.27	Wheelchair	2017
23	F	USA	64.30	133.12	Wheelchair	2017
39	F	GBR	64.35	133.17	Wheelchair	2017
32	F	VEN	93.53	203.27	Wheelchair	2017
47	M	ALB	116.02	246.30	Wheelchair	2017
51	F	USA	166.32	340.32	Wheelchair	2017
38	M	USA	187.57	360.88	Wheelchair	2017
51	M	ECU	171.88	377.72	Wheelchair	2017
58	M	USA	186.18	412.48	Wheelchair	2017
35	F	VEN	206.10	423.65	Wheelchair	2017
32	F	VEN	208.60	441.50	Wheelchair	2017
64	M	USA	220.63	469.65	Wheelchair	2017

ISBN: 9780170425711

ISBN: 9780170425711

Answers

Revision (pp. 7–15)

Types of variables (pp. 7)

Variable	Type of variable	Suitable for inference?
Favourite ice-cream flavour	Descriptive	✗
Height	Continuous	✓
Number of apps on your phone	Discrete	✗
Neck circumference	Continuous	✓
Shoe size	Discrete	✗
Number of siblings	Discrete	✗
Size of your house (m^2)	Continuous	✓
Number of countries you have visited	Discrete	✗
Hair colour	Descriptive	✗
Brand of shoe	Descriptive	✗
Distance to the nearest supermarket	Continuous	✓
Length of longest finger	Continuous	✓
Weight of school bag	Continuous	✓
Amount of water you drink today	Continuous	✓
Age	Continuous however could be considered discrete if measured in years or months etc.	✗

Box plots (pp. 8–9)

1

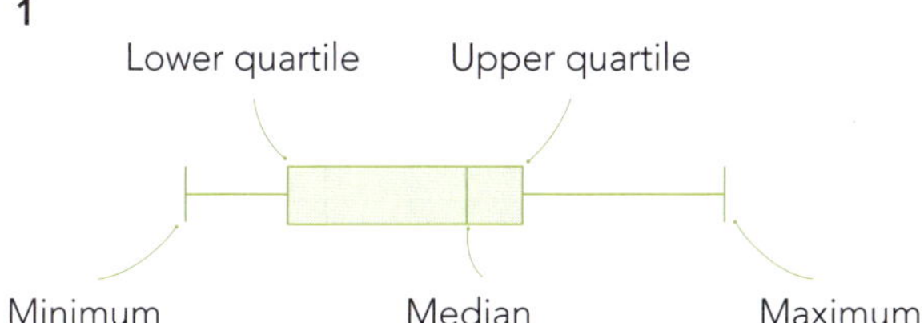

2 **Boys**:

Minimum 5 — Lower quartile 10
Median 13 — Upper quartile 17
Maximum 21

Girls:

Minimum 3 — Lower quartile 8
Median 14.5 — Upper quartile 19
Maximum 23

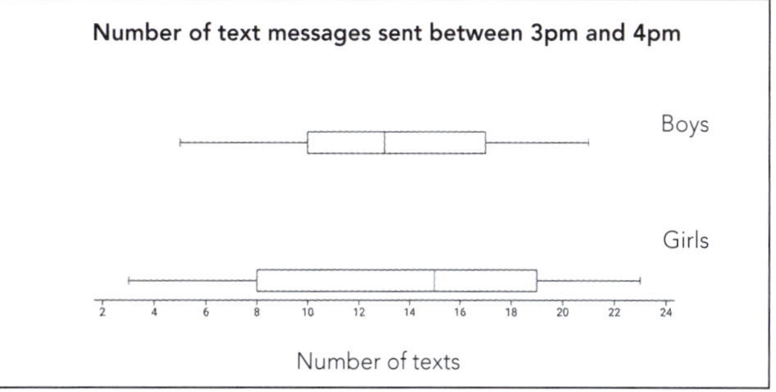

3

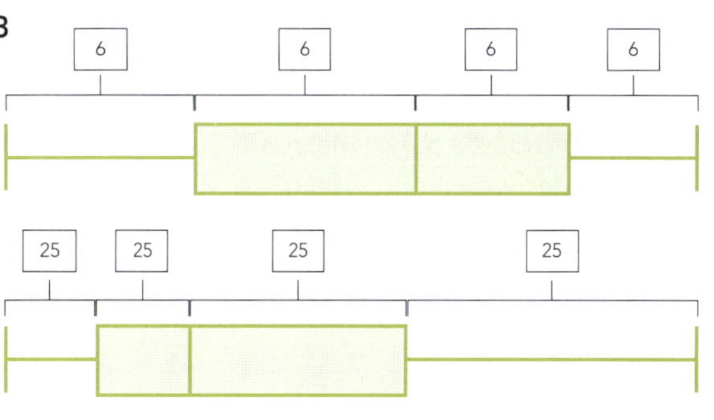

4

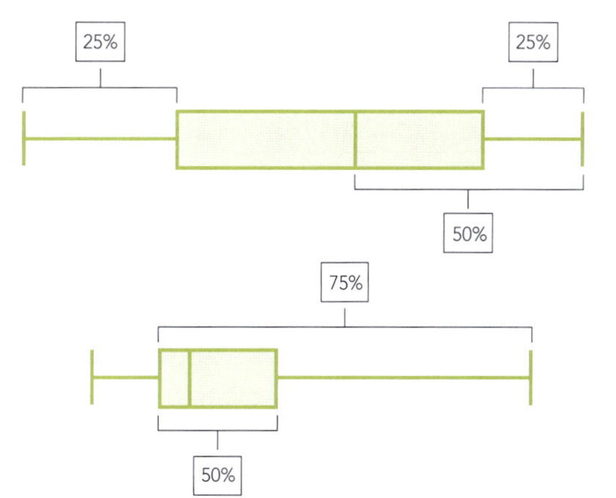

5

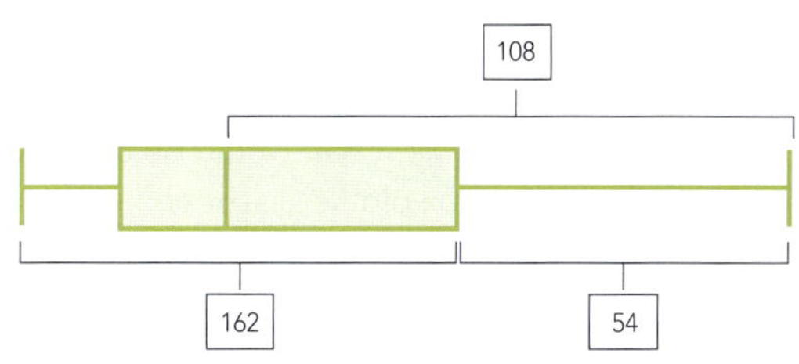

Shapes of distributions (p. 10)

1 Name: Bell shaped distribution
Description: Symmetrical and bell-shaped

2 Name: Left skew
Description: The tail is on the left-hand side.

3 Name: Uniform
Description: Looks like a box.

4 Name: Bimodal
Description: There are two peaks.

ISBN: 9780170425711

5 Name: Right skewed
Description: The tail is on the right-hand side.

6 Name: Triangular
Description: Symmetrical and triangular shape.

7 Name: Irregular
Description: No real pattern.

Use of medians and means (pp. 11–15)

1 a Median = 8 Mean = 10.666
It would be appropriate to use the median because the data set is not symmetrical – the 29 pulls the mean up.

b Median = 19 Mean = 20
It would be appropriate to use either median or mean because the data set is reasonably symmetrical.

c Median = 9 Mean = 10.92
It would be appropriate to use the median because the data set is not symmetrical – the 40 pulls up the mean.

2 Boys: Median = 8.5 Mean = 8.56
Girls: Median = 4 Mean = 6

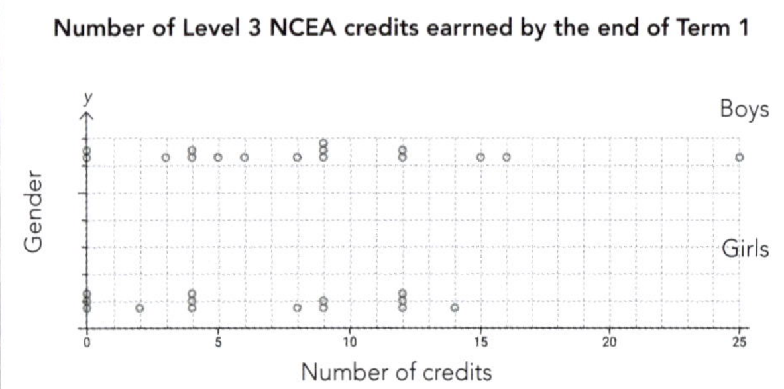

For the boys' data, the median is more appropriate to use because the spread is not symmetrical.
For the girls' data, the median or mean is appropriate to use because the spread is reasonably symmetrical.

3 a Either median or mean would be suitable because the dot plot is reasonably symmetrical.

b The median would be suitable because the dot plot is skewed to the left and not symmetrical.

c Either median or mean would be suitable because the box plot is reasonably symmetrical.

d The median would be suitable because the box plot is slightly skewed.

e The median would be suitable because the box plot is slightly skewed.

f Appropriate measure(s) of centre for a symmetrical distribution: either median or mean.

g Appropriate measure(s) of centre for an asymmetrical distribution: the median.

Sampling variability (pp. 16–18)

1 These are an example of what samples might be like; your samples will be different. Check with your teacher.

Random number	Kea	Mass (g)
17	Taan	1000
14	Moritz	900
27	Aroha	700
12	Anthea	750
9	Hoheria	780
5	Lorraine	850
15	Marie	770
7	Sage	950
18	Alwyn	800

Median = 800 g

Random number	Kea	Mass (g)
9	Hoheria	780
2	Sneaky	840
5	Lorraine	850
19	Yin	680
17	Taan	1000
15	Marie	770
21	Zeal	650
24	Roxanne	690
12	Anthea	750

Median = 770 g

Random number	Kea	Mass (g)
12	Anthea	750
30	Handlebars	915
25	Anna	730
6	Christmas	850
23	Gizo	860
22	Swazi	675
18	Alwyn	800
17	Taan	1000
26	Juno	900

Median = 850 g

The medians are all different; each sample is likely to produce different medians.

2 These are an example of what your samples might be like; your samples will be different. Check with your teacher.

ISBN: 9780170425711

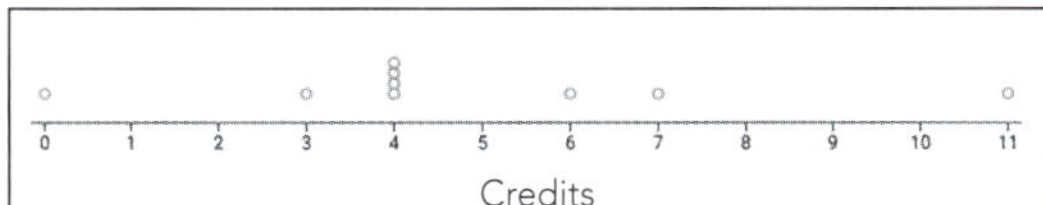

Median = 4 credits

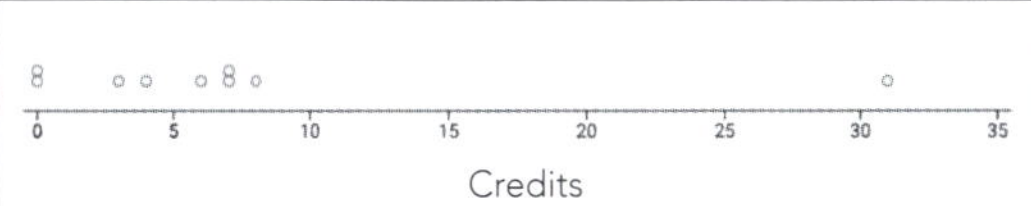

Median = 6 credits

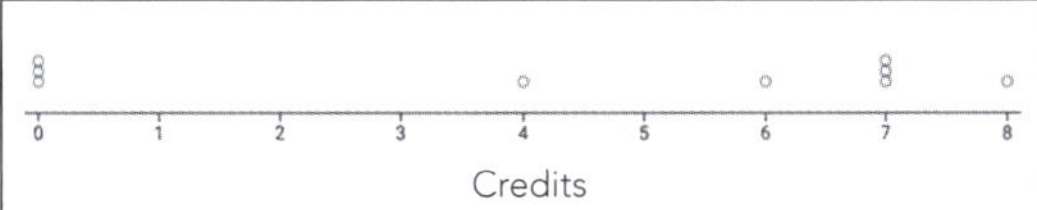

Median = 6 credits

In this case, the medians are not all different; each sample may or may not produce the same median.

Resampling (pp. 19–21)

1

Resample 2	
Number	**Mass (g)**
5	880
6	1010
6	1010
1	860
4	820
6	1010

Median = 945 g

These are an example of what your samples might be like; your samples will be different. Check with your teacher.

Resample 3	
Number	**Mass (g)**
6	1010
5	880
5	880
6	1010
3	840
2	930

Median = 905 g

2 Median of the sample = 174.5 cm

These are an example of what your samples might be like; your samples will be different. Check with your teacher.

Resample 1	
Number	**Arm span (cm)**
1	180
2	175
10	181
4	187
9	153
5	162
4	187
2	175
5	162
2	175

Median = 175 cm

Resample 2	
Number	**Arm span (cm)**
10	181
6	174
3	170
8	185
8	185
1	180
6	174
1	180
9	153
4	187

Median = 180 cm

Resample 3	
Number	**Arm span (cm)**
5	162
4	187
3	170
7	173
1	180
9	153
8	185
1	180
5	162
5	162

Median = 171.5 cm

Resample 4	
Number	**Arm span (cm)**
7	173
5	162
9	153
1	180
9	153
4	187
8	185
3	170
1	180
2	175

Median = 174 cm

The difference between two medians (or means) (pp. 22–29)

1 Male median: 175cm Females median: 165cm
The male median arm span is larger than the female in the sample.
For every resample we subtract the female resample median from the male resample (Male – Female).

Resample 1

Males

Number	Student	Arm span (cm)
1	Bruce	180
1	Bruce	180
4	Hemi	187
2	James	175
1	Bruce	180

Median = 180 cm

Females

Number	Student	Arm span (cm)
3	Elaine	162
5	Eleanor	167
2	Charlie	154
1	Maya	165
3	Elaine	162

Median = 162 cm

Calculate the difference in the medians. 180 – 162 = 18 cm
What does this mean for this resample? **This suggests that the male's arm spans at Paradise High School are on average 18 cm larger than the female's at Paradise High School.**
In this resample the difference between the medians was positive. This means that in this resample, the median arm span of the males was larger than that of the females.

ISBN: 9780170425711

Resample 2

Males

Number	Student	Arm span (cm)
5	Trevor	162
3	Spencer	153
1	Bruce	180
3	Spencer	153
4	Hemi	187

Median = 162 cm

Females

Number	Student	Arm span (cm)
1	Maya	165
2	Charlie	154
1	Maya	165
5	Eleanor	167
4	Jane	171

Median = 165 cm

Calculate the difference in the medians. **162 – 165 = –3 cm**

What does this mean for this resample? **This suggests that the males' arm spans at Paradise High School are on average 3 cm smaller than the female's at Paradise High School.**

In this resample the difference between the medians was **negative**. This means that in this resample, the median arm span of the **females** was larger than that of the **males**.

Resample 3

Males

Number	Student	Arm span (cm)
5	Trevor	162
4	Hemi	187
2	James	175
5	Trevor	162
3	Spencer	153

Median = 162 cm

Females

Number	Student	Arm span (cm)
2	Charlie	154
5	Eleanor	167
3	Elaine	162
4	Jane	171
2	Charlie	154

Median = 162 cm

Calculate the difference in the medians. **162 – 162 = 0 cm**

What does this mean for this resample? **This suggests that the males' arm spans at Paradise High School are on average the same as female's.**

In this resample the difference between the medians was **zero**. This means that in this resample the median arm span of the **females was the same as the males**.

2 Male median: **870 g** Females median: **854 g**

In the resamples I will subtract the female medians from the male medians.

These are an example of what your samples might be like; your samples will be different. Check with your teacher.

Resample 1

Males

Random number	Kea ID	Mass (g)
4	Baggins	820
5	Bandit	880
4	Baggins	820
1	Daffy	860
1	Daffy	860
5	Bandit	880

Median = 860 g

Females

Random number	Kea ID	Mass (g)
3	Cassidy	790
4	Creeky	868
1	Beryl	840
6	Sage	950
1	Beryl	840
6	Sage	950

Median = 854 g

Difference between the medians = 6 g

Resample 2

Males

Random number	Kea ID	Mass (g)
3	Herbert	840
3	Herbert	840
6	Allsorts	1010
5	Bandit	880
2	Cheezel	930
1	Daffy	860

Median = 870 g

Females

Random number	Kea ID	Mass (g)
4	Creeky	868
6	Sage	950
3	Cassidy	790
5	Trixy	900
3	Cassidy	790
2	Xena	830

Median = 849 g

Difference between the medians = 21 g

Resample 3

Males

Random number	Kea ID	Mass (g)
5	Bandit	880
6	Allsorts	1010
5	Bandit	880
5	Bandit	880
2	Cheezel	930
3	Herbert	840

Median = 880 g

Females

Random number	Kea ID	Mass (g)
5	Trixy	900
4	Creeky	868
4	Creeky	868
2	Xena	830
1	Beryl	840
3	Cassidy	790

Median = 854 g

Difference between the medians = 26 g

Resample 4

Males

Random number	Kea ID	Mass (g)
1	Daffy	860
6	Allsorts	1010
5	Bandit	880
6	Allsorts	1010
5	Bandit	880
2	Cheezel	930

Median = 905 g

Females

Random number	Kea ID	Mass (g)
5	Trixy	900
4	Creeky	868
2	Xena	830
5	Trixy	900
3	Cassidy	790
4	Creeky	868

Median = 868 g

Difference between the medians = 37 g

Resample 5

Males

Random number	Kea ID	Mass (g)
6	Allsorts	1010
3	Herbert	840
1	Daffy	860
1	Daffy	860
4	Baggins	820
2	Cheezel	930

Median = 860 g

Females

Random number	Kea ID	Mass (g)
6	Sage	950
2	Xena	830
3	Cassidy	790
6	Sage	950
6	Sage	950
2	Zena	860

Median = 890 g

Difference between the medians = –30 g

ISBN: 9780170425711

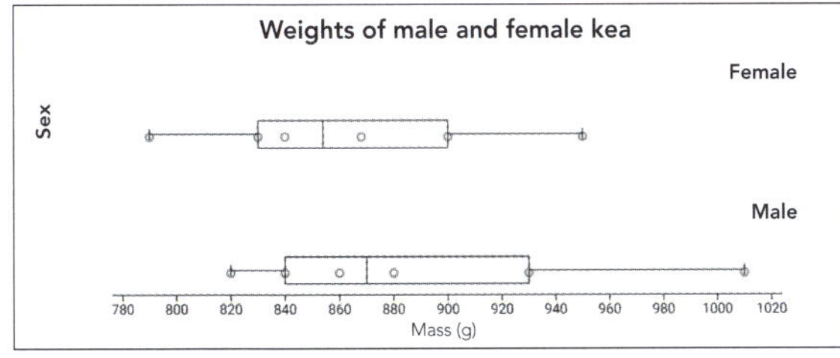

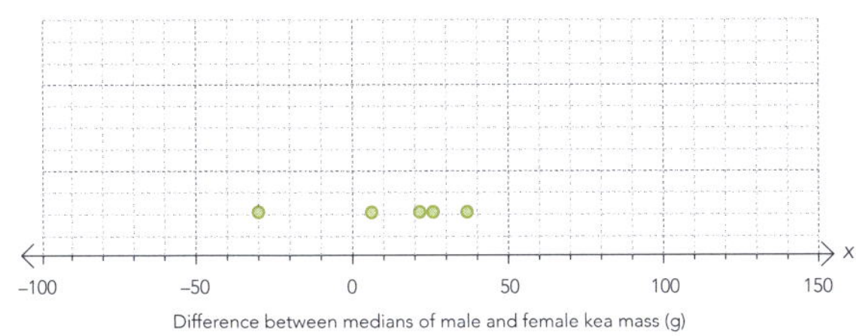

Computer program output (p. 26)

1 In this graph we are comparing the means rather than the medians so there is no boxplot.

2 Firstly, because in the second graph we are comparing the mean rather than the medians. Secondly, each time you generate a confidence interval you will get different values. This is due to sampling variability.

3 Because in the original sample the males crown length is larger than the females crown length.

Does sample size make a difference? (pp. 28–29)

Sample size	Upper limit of confidence interval	Lower limit of confidence interval	Difference = width of confidence interval
10	260	**82**	178
40	**202.5**	50	**152.5**
1000	**200**	**97.5**	**102.5**

The width of the confidence interval becomes **smaller/narrower** as the sample size **increases**.

Setting up your investigation (pp. 30–34)

1 Select suitable data (p. 31)

1 a

Group one	Group two	Variable
Male	Female	Beak length (mm)
Male	Female	Crown length (mm)

2 You could compare the weight difference of the the age groups, however you would need to narrow the data set to just two age groups rather than all of them.

2 Pose an appropriate question, using suitable data (pp. 32–33)

1 a Good question? No
Rewritten question: What is the difference between **the median** number of daily sunshine hours in Auckland and Nelson?

b Good question? Yes

c Good question? No
Rewritten question: What is the difference between **the median** arm span (cm) of **males** and **females** at Paradise High School?

2 a What is the difference between the median height (cm) of male and female students at Paradise High School?

b What is the difference between the median bag weight (kg) of Year 11 and Year 9 students at Paradise High School?

Describing and comparing sample distributions (pp. 35–39)

1 **Centre:**
I notice that the median arm span for males at Paradise High School is 13.5 cm greater than the median span for females. This means that in the sample, males have longer arm spans on average than females.

Shape:
I notice that the distribution of the females' arm spans tends to be skewed to the left. The shape of the distribution of the males' arm spans tends towards a bell shaped distribution. If we had a larger sample for the females, it may tend towards a bell shaped distribution as well.

Spread and overlap:
The range of the males' arm spans is larger than the range of the females' arm spans. The interquartile ranges are similar in size, although the middle 50% of the males' data is further to the right than the middle 50% of the females' arm spans. 75% of the males' arm spans are larger than 75% of the females' arm spans. Almost all of the males' data is larger than the females' median arm span.

Unusual features:
I notice that there is a male with an unusually short arm span of 98 cm. There is also a female with a very short arm span of 100 cm. While these are possible, there may have been an error in measuring or recording of these people. There is also a male with a very long arm span of 212 cm. While this seems very long for a Year 13 student, it is possible.

2 I notice that the median number of Level 2 NCEA credits for females is 1.5 credits more than the males' median. This is not a large difference. I notice that the shapes of the distributions for both the male and female numbers of credits tend to be irregular.

ISBN: 9780170425711

I notice that the range for the males (138 credits) is larger than that for the females (107 credits), and it is further to the left. This is because a number of male students haven't received many credits. One male student hasn't received any. This may be an international student who arrived late in the year or one who isn't sitting NCEA.
The interquartile range for males is 41.5 credits, this is larger than the females (26 credits). The middle 50% of males credits is further to the left than the middle 50% of females. The distribution of the number of credits males have received is more spread out than that for females, which indicates there is more variability in the number of credits received by this sample for males.

Making a formal inference about the difference between medians (pp. 40–43)

1 We can be reasonably confident that for the populations of **people with no qualifications and those with university qualifications** from **New Zealand**, the median **weekly income ($)** will be between **$300** and **$550** greater for those with university qualifications than those with **no qualifications**.
As zero **isn't** included in the bootstrapped confidence interval, we **can** conclude that there is a difference in the median **weekly income ($)** between those **with university qualifications** and those with no qualifications.

2 We can be reasonably confident that **for the populations of male and female students from Paradise High School, the arm spans will be between 11 cm and 21 cm greater for male students than female students.**
As zero isn't included in the bootstrapped confidence interval, we can conclude that there is a difference in the arm span between male and female students from Paradise High School.

3 We can be reasonably confident that **for the populations of Douglas fir and Pinus radiata trees from a foresty block, the median height will be between 0.67 m shorter and 1.02 m taller for Douglas fir than Pinus radiata.**
As zero is included in the bootstrapped confidence interval, we cannot conclude that there is a difference in the height (m) between Pinus radiata and Douglas fir in this forestry block.

Making a formal inference about the difference between means (pp. 44–46)

1 We can be reasonably confident that for the populations of **those with no qualifications and with university qualifications** from **New Zealand**, the mean **hours worked per week** will be between **3.65 hours** less and **4.879 hours** greater for those with no qualifications than those with **university qualifications**.
As zero **is** included in the bootstrapped confidence interval, we **cannot** conclude that there **is** a difference in the mean **number of hours worked** between those **with no qualifications** and those with university qualifications.

2 We can be reasonably confident that for the **populations of Douglas fir and Pinus radiata trees from a foresty block, the girth will be between 25.8 cm and 30.95 cm greater for Douglas fir than Pinus radiata.**
As zero isn't included in the bootstrapped confidence interval, we can conclude that there is a difference in the girth (cm) between Pinus radiata and Douglas fir in this forestry block.

3 We can be reasonably confident that **for the populations of male and female students from Paradise High School, the mean bag weights will be between 0.3986 kg lighter and 1.2357 kg heavier for male students than female students.**
As zero is included in the bootstrapped confidence interval, we cannot conclude that there is a difference in the bag weights between male and female students from Paradise High School.

Writing about statistical inference (pp. 47–48)

1 I chose to use the median rather than the mean because the median **isn't affected because the data had no unusual points.**

2 In the graphs we observe that 75% of the money that male students spent on formal wear is less than 50% of the money female students spent on formal wear. This **means** that in the sample, female students on average **do** spend a greater sum of money on formal wear than male students.

3 I **can** make the call that the median weight for female kiwi was 2.9015 kg and for male kiwi was 2.246 kg with a difference of 0.6555 kg. This indicates that for this sample, females **are** on average heavier than males.

4 In this sample 75% of the female athlete body fat percentages are greater than 75% of the male athlete body fat percentages. This **means** that for this sample, the females have higher body fat percentages on average than males.

5 By looking at the data for weights, researchers might also be able to **suggest** whether or not the kiwi have access to enough food.

6 I notice that in the sample the interquartile range (IQR) of the weight of female kiwi (0.507 kg) is

ISBN: 9780170425711

greater than the IQR of male kiwi (0.18 kg). This **means** that the middle 50% of female kiwi weights are more spread that the middle 50% of male kiwi weights.

7 The spread in the middle 50% for the males is wider than the middle 50% for the females, this **means** that the males have more variation in the sample data compared with females.

8 If I were to use the mean as an average for sample distributions that are not symmetrical, it **could** affect my investigation as the mean **can** be influenced by extreme values.

Pick the errors (p. 49)

1 The difference in sample size doesn't determine whether the sample is symmetrically distributed or not.

2 We have no way of knowing what the shape of the population data will be. You may be able to suggest a shape for the population with justification.

3 The median length of sharks is not 5–9.5 cm, as this would be a very small shark. The difference in the median length will be between 5 and 9.5 cm.

4 The confidence intervals are making a suggestion about the population, not the sample.

5 The interquartile range is a value and therefore doesn't have a position. The middle 50% of the data can be further to the right or left.

6 More data may give us a more distinct shape. This is not due to sampling variability.

7 The use of higher is ambiguous. The forwards range is 11 kg wider.

8 The confidence interval gives us values that we expect the difference in the median to be; we cannot say that the difference will always be between these values. Also missing 'median' in this sentence.

What is the population? (p. 50)

1 **d** All babies born in Takapuna Hospital in 1982.

2 **c** All credit card transactions in Hamilton on Boxing Day on the year of the sample.

3 **a** The time it takes all Paradise High School students to get to school.

4 **c** The amount of chocolate consumed by each Waiheke Island occupant on Christmas Day.

5 **b** Hours of television watched during the holidays by all Year 11 students from Paradise High School.

Practice tasks (pp. 54–66)

Practice task one (pp. 54–56)

1 The question has a yes/no answer. The question should ask what the difference is between the median delivery times, not whether one is longer than the other.
A better question would be: What is the difference between the median delivery times (minutes) of Pizzarella and Crusty?

2 The variable is incorrect – it should be the pizza delivery time rather than the time taken to make the pizza.

3 Population hasn't been identified.

4 Crusty's median delivery time is **25** minutes (the mean was 24.779, not the median).

5 The median delivery time for Pizzarella is higher than the upper quartile for Crusty. This means that 75% of Crusty's deliveries take **less time** than half of Pizzarella's.

6 I'm confident that for these pizza delivery companies the difference between the medians is between **1** and **8** minutes. (The difference between the medians of the sample was used rather than the lower limit of the confidence interval.)

7 This means that Crusty's pizza delivery is always going to be faster than Pizzarella's. **This statement is incorrect. We could conclude that the median time is likely to be higher. We cannot conclude that all deliveries will be faster from Crusty than Pizzarella.**

8 If I did this project again, I would probably get different displays and sample statistics because the companies might have changed owners. **Due to sampling variability, there are likely to be different results regardless. The change in ownership would not be the main reason that results would differ.**

Practice task two (pp. 57–60)

Question and purpose:

What is the difference between the median distance (km) covered by defensive players and offensive players in their weekend hockey game from Suburbanville?

Those who might find this report useful would be sports scientists and game officials (coaches, trainers, etc.) who can use the findings to help establish goals for their players. The information provided could be used to specify fitness requirements for those who wish to play in representative teams. This information might also help someone who is trying to decide the position for a player, because some positions require a higher level of anaerobic fitness than others (https://www.fieldhockeyreview.com/complete-field-hockey-positions-overview/).

ISBN: 9780170425711

Define your variables and population:
The primary responsibility of defensive players is to prevent the opponents from scoring goals. The positions of these players are left and right halfback, two fullbacks and goalkeeper. The role of offensive players is to score goals against the opposition. The positions of these players are left and right wing and the centre forward. There are three players unaccounted for. These are the inside-left and inside-right (inners) positions, and the centre half. (https://www.teamusa.org/usa-field-hockey/usa-field-hockey-101). These players are considered to have dual roles and form the midfield, which is both offensive and defensive. It is unclear whether they have been considered offence or defence in this data set.

The distance covered is in kilometres. Although we are assuming this is for one game (70 minutes), it may in fact be less than 70 minutes as we do not have any information on how long each of these players were on the field for (http://www.fih.ch/media/12236728/fih-rules-of-hockey-2017.pdf). We also cannot assume that they have run these distances, as there would be periods of time when players would walk.

Since the data was collected over a weekend in Suburbanville, we can consider the population to be all adult hockey players from Suburbanville who play during the weekend.

Discuss and compare sample distributions:
The median distance covered by the defensive adult hockey players in this sample is 5.95 km. This is 2.95 km less than that for the offensive hockey players, who covered a median distance of 8.9 km in a game. The defensive hockey players' data is more spread out than that for the offensive players. The range for the defensive players is 8.2 km while for the offensive players it is 3.4 km. This is partly because of an unusual point for one of the defensive players. This player covered only a distance of 0.6 km in a game. This player likely to be the goalkeeper. However it could also be a player who was substituted and played for only a short time in the game. It is unusual because it is around 4 km less than any other player in the sample.

The IQR of distance covered of the offensive players is 1.02 km whereas for the defence it is 1.55 km. This indicates more variability in the distance covered by the middle 50% of defensive players in the sample.

Of significance is the fact that all of the offensive players have covered a greater distance in their game than 75% of the defensive players. The minimum distance covered by an offensive player is 6.81 km and the upper quartile for the defensive players is 6.72 km. This means that the majority of the offensive hockey players covered a greater distance than the majority of the defensive players.

The shapes of both samples tend towards irregular, but this may be due to the size of the samples being relatively small.

Make a formal inference:
We can be reasonably confident that for these populations of offensive and defensive players from Suburbanville, the median distance covered in a hockey game will be between 2.57 km and 3.53 km further for offensive players than for defensive players.

As zero isn't included in the bootstrapped confidence interval, we can make the call that there is a difference in the median distance covered in a hockey match between offensive and defensive players in Suburbanville.

Discuss sampling variability:
I am assuming that the original data set is representative of the population of all adult hockey players in Suburbanville. If I obtained another sample, due to sampling variability the display and sample statistics would probably be different from those above. However, because there is a significant difference between the two positional distances covered in a game, I am confident that I would come to the same conclusion.

Conclusion:
Based on these samples I can make the call that adult offensive hockey players cover a greater distance than adult defensive hockey players from Suburbanville. I can make this call because the bootstrapped confidence interval is positive.

This conclusion could be useful for players who are wanting to select their position in a team. It gives them information about which positions require greater aerobic fitness. It also suggests that offensive players need to be able to run on average 2–3 km further than defensive players. This knowledge may benefit those developing fitness training programmes for the players.

Reflection, limitations and further investigation:
For this report I used the median as a measure of average. While the samples are not symmetrical the sample means are similar to the sample medians, for this reason the means could also have been used.

As this data set was only taken from a weekend in Suburbanville, there are a number of limitations to this study. We are not provided with information on gender, age, level of competition, weather, etc. All of these aspects could have a bearing on the data collected and therefore who the findings could be applied to.

Practice task three (pp. 61–66)
Question:
What is the difference between the median finishing

ISBN: 9780170425711

times (minutes) of handcycle and wheelchair competitors from the 2016 and 2017 New York Marathon? The sample for this report was provided by https://www.tcsnycmarathon.org/.

Purpose:
Those who might find this report useful would be marathon organisers. It would help establish the need for staggered starts and give an indication about when they would need staff out on the course to aid competitors. This is likely to be earlier than runners as than able-bodied competitors.

Those using handcycles and wheelchairs are likely to travel faster than runners. Therefore, if they all started together, the staff supporting those using handcycles and wheelchairs would need to be available earlier than those supporting runners.

Define the variables and population:
Competitors were either in a wheelchair or competing in a handcycle. One of the main differences between these is that a handcycle has gears and a brake, whereas racing wheelchairs do not. It is speculated that a handcycle is easier for disabled athletes to use and is less stressful on the body (https://www.nytimes.com/2006/11/02/sports/sportsspecial/02handcycle.html). This allows those with less athletic experience and more restrictive disabilities to compete. This may result in a greater variation (range) of results within the handcycle category.

The marathon organisers also restrict hand cyclists by not allowing them to participate if their anticipated finish time is faster than 1 hour and 35 minutes. (https://www.tcsnycmarathon.org/plan-your-race/run-in-2018/how-to-apply-athlete-with-disability-awd). This and the lack of prize money for hand cyclist indicates that they are allowed entry to participate, rather than to compete.

The overall time is the official time it took for each participant to complete the 42.195 km race.
The population is all athletes who competed in the New York Marathon in either 2016 or 2017 in a wheelchair or handcycle.

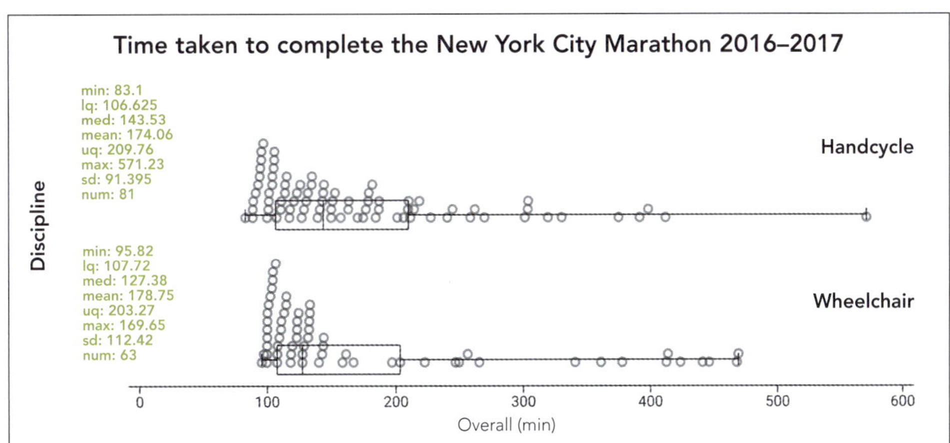

The median finish time for the wheelchair participants was 16.15 minutes faster than the median finish time of the handcycle participants. The middle 50% of the wheelchair times is slightly less spread out and is within the 50% of the handcycle finish times. This means there was slightly more variation in the middle 50% of the handcycle times than the wheelchair marathon times.

The range for the handcycle times (488.13 min) is longer than the range for the wheel chair times (373.83 min). This is partly due to an unusually long hand cyclist's time of 571.23 minutes. This time is 159.5 minutes lower than the next closest handcycle time.

The shape of both data sets tends towards a right skew. For both groups the slowest 25% took much longer times than most, probably reflecting varying degrees of disability.

There was prize money for the wheelchair competitors, but not for those using handcycles. This is probably the reason that the finish times for the fastest 50% of hand wheelchair competitors covers a much narrower interval than that for the hand cyclists. This might also be due to the fact that hand cyclists were not allowed to enter if their times had been less than 1 hour and 35 minutes.

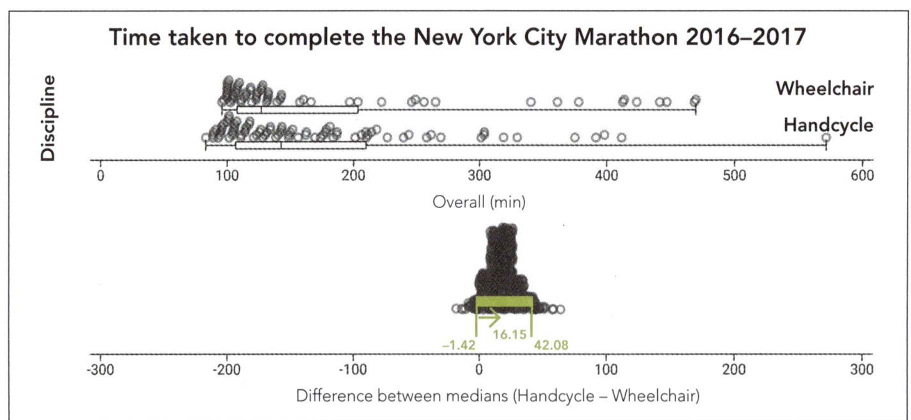

We can be reasonably confident that for these populations of New York Marathon handcycle and wheelchair participants, the median marathon time will be between 1.42 minutes less and 42.08 minutes more for handcycle users than wheelchair athletes.

As zero is included in the bootstrapped confidence interval, we cannot make the call that there is a difference between the median handcycle and wheelchair participants marathon time.

For this report I used the median as a measure of the average. I felt this was appropriate as both the sample distributions were skewed. The mean is not an appropriate measure of centre for a right-skewed distribution because it would be influenced by the wide spread in the top 25% of times. I am assuming that the original data set is representative of all wheelchair and handcycle participants in the 2016/17 New York Marathon. If I obtained another sample, due to the sampling variability the display and sample statistics would probably be different.

Conclusion:

Based on these samples I cannot make the call that the hand cyclists' median marathon finish time will be on average faster than that for athletes in a wheelchair. It is possible that the restrictions on hand cyclists are intentional so that they finish at a similar time to wheelchair participants. This may be so that the course officials are required to be in their positions no earlier for one discipline than the other.

This report could also be used to confirm the restriction of 1 hour and 35 minutes for hand cyclists. This seems an appropriate cut-off compared with the majority of the wheelchair participants.

Reflection, limitations and further investigation:

It would be interesting to investigate the effect of the restriction on the handcyclists' finishing times. This could be done by randomly dividing the hand cyclists into two groups, one of which had the restriction, and one which didn't. Alternatively the organisers could remove the restriction for one year and compare the data. This may not produce results that could be compared as easily, due to such factors as different weather conditions, technological improvements in the cycles, etc.

It would also be interesting to investigate the effect of offering prize money for those in wheelchairs, but not for those using handcycles. This could be done by offering the same prize money for both groups for one year, and comparing the results with those for the previous years. Again, factors such as different weather condition and technological improvements may have to be accounted for.

This study used results only from 2016 and 2017. These results may not be comparable with those from other years, or to future results, due to changes in technology.

It also used results from only the New York Marathon. It would be interesting to compare these with others, such as the Boston Marathon.

ISBN: 9780170425711